Compression Mode

Compression Mode

The Edge of Sensibility

Stephen Kennedy

BLOOMSBURY ACADEMIC
NEW YORK • LONDON • OXFORD • NEW DELHI • SYDNEY

BLOOMSBURY ACADEMIC
Bloomsbury Publishing Inc
1385 Broadway, New York, NY 10018, USA
50 Bedford Square, London, WC1B 3DP, UK
29 Earlsfort Terrace, Dublin 2, Ireland

BLOOMSBURY, BLOOMSBURY ACADEMIC and the Diana logo are trademarks of Bloomsbury Publishing Plc

First published in the United States of America 2025

Cover design by Louise Dugdale
Cover image: 3D SLAM scan of Cleopatra's Needle, London, top view
Simon Withers, Captivate Heritage Laboratory

Library of Congress Cataloging-in-Publication Data
Names: Kennedy, Stephen, author.
Title: Compression mode : the edge of sensibility / Stephen Kennedy.
Description: [1.] | New York : Bloomsbury Academic, 2025. | Includes bibliographical references.
Identifiers: LCCN 2024025140 (print) | LCCN 2024025141 (ebook) | ISBN 9781501369346 (hardback) | ISBN 9781501369353 (paperback) | ISBN 9781501369360 (ebook) | ISBN 9781501369377 (pdf)
Subjects: LCSH: Music—Philosophy and aesthetics. | Data compression (Computer science) | MP3 (Audio coding standard) | Sound (Philosophy) | Noise (Philosophy)
Classification: LCC ML3800 .K33 2025 (print) | LCC ML3800 (ebook) | DDC 781.2/3—dc23/eng/20240717
LC record available at https://lccn.loc.gov/2024025140
LC ebook record available at https://lccn.loc.gov/2024025141

ISBN: HB: 978-1-5013-6934-6
ePDF: 978-1-5013-6937-7
eBook: 978-1-5013-6936-0

Typeset by Newgen KnowledgeWorks Pvt. Ltd., Chennai, India

To find out more about our authors and books visit www.bloomsbury.com and sign up for our newsletters.

For Gemma, Lori, and Erin

Contents

Preface

Compression Mode is a book about space-time. Setting out from a fusion of dimensions, it examines how together the fullness of the universe and the figure of compression can be understood as a singularity. Compression is presented not only as a digital process enacted through computing to reduce and accelerate, but also as a wider economic and political phenomenon that impacts on ecologies of waste, diversity, and social inclusivity.

The abiding ambition of this volume is to challenge the received wisdom that reason and logic, as primarily human activities, serve to deliver a coherent resolution to universal concerns, with order ultimately triumphing over chaos. As a counterpoint to this an alternative is proposed, arguing instead that reason and logic serve to impose certain very specific *modal* patterns on a world not entirely conducive to comprehensive or conclusive ordering. This ordering (as a mode of compression), such as it is, occurs through a process of mediation and representation and is communicated via an evolving set of technological standards. So, the term 'compression' as it is deployed here is understood simultaneously as the imposition and the communication of a kind of stable (relatively) and perceptible set of norms. In contemporary settings, however, such an imposition is no longer substantive, and does not satisfactorily account for the universe or the place of humans in it. To better serve our understanding of the circumstances we find ourselves in, a reassessment of the degrees of complexity that are present in subject/object, analogue/digital, human/posthuman/more than human relations, is necessary.

Embracing the *multiple* and the philosophical imperative of *assemblage*, *Compression Mode* collects references, inspirations, and points of interest as it goes – sometimes coherently while at others taking vectoral tangents – always energized and dynamic, sometimes settling, hesitating to rest and 'make sense'. Through a process of orchestration, the various players are mixed and composed, sometimes tonally and sometimes shaped by atonal serialism, at other times improvised.

Across the pages that follow, mathematical patterns form a protean complex with compression serving to create moments of order and clarity. Ever present are the unthinking, mindlessness of nature, self-organization and pattern formation, noise, waste, and repetition as process. The approach chosen presents as a particular kind of repetition, where difference is enacted through resistance, and where noise names an unfolding and unstable visceral affect that is distinct from the critical distance of the intellect. Whereas the latter resolves to extricate itself from uncertainty, the former embraces it and rather than being a secondary emotional form of intellect, is a fuller, more complex, and expressive form.

Through music as art and art as music, the coming to dominance of the visual and the linguistic turn, and finally a return to the acoustic and to noise, the narrative unfolds as a non-linear topology formed from regions rather that points as new and strange geometries, spherical and harmonious, come to serve as our 'measure'.

The grain of the voice, intuition, the collapse of subject/object and of narrator and story, where story is itself a compressed form of reality, combine as they do in the work of Borges to both reduce and expand the complexity of the event horizon.

The compression of data is more than a metaphor. It is a practice that constitutes 'political economy by design' and as such it serves to marginalize and exclude. The practice resonates with the rise of populism and the prevalence of the 'lowest common denominator', the pushing of difference to the margins (while simultaneously celebrating its triumph), and the intensifying of nationalism and retrenchment.

Finally, the digitization of life in all its forms continues unabated. Reconciled and described using data rendered into code, life as both nature and culture is increasingly organized into ordered patterns. When the word data is used in this context, it generally refers to information that has/can be transformed and compressed into language for computers to process. If it is to be understood as more than this, however, data must be recognized as having a primordial presence that has always existed, as measurable and immeasurable quanta, it is nothing new. It is rather the sum total of all potentiality: all the information that exists, accepted and contested in a universe where order always breaks down. With this firmly in mind, we enter the library of Babel and begin our quest.

Introduction

Digitization is the latest in a long line of symbolic 'technologies' that have sought to impose order and a degree of comprehensibility on a wild and unruly universe. Its predecessors comprise a multitude of practices and linguistic models that include mathematics, geometry, and musical notation. A journey through the history of these 'technologies' shows how the drive to compress (that is central to all of them) comes at a cost commonly characterized by exclusion, the entrenchment of bias, and ideological dogma. This, as will become clear, is testament to the tendency towards superficiality that such practices all too often display. If this is not to continue into the digital age, then these symbolic practices need to be understood as processes and the details of their unfolding must be rethought and reconstituted to include contrasting and relative notions of expansion and inclusion, even when this generates a degree of uncertainty. It is not the case that compression is a negative practice and expansion a positive one, but rather that the interoperability between the two needs to be properly understood and articulated across all modes of experiencing.

The compression and expansion of time and space and their representation, as means of understanding and exerting control over the conditions of our shared existence as well as our individual experience (this distinction is important and will be expanded upon further in this volume) of the world, are the key areas of concern in what follows. Different modes of understanding and exerting control might be a matter of science and technology, but they are also very much the concern of philosophy and aesthetics. Why is this important? Because all incidences of compression or expansion of time and space – whether enacted by humans explicitly (cultural phenomena), or as purely 'natural' phenomena (and these two options are not mutually exclusive), or technological and mathematical – change the constitution of objects and subjects, relative to one another in a complex universe of interconnectivity. Currently, the focus of such change is on environmental matters, global political economy, and the avoidance of catastrophe, yet there seem to be few answers to these challenges that adequately go beyond current received wisdom. Minor adjustments to established arrangements and privileges are insufficient. What follows will aim to introduce several complex and specialized positions that when combined can be simultaneously compressed and expanded into forms that communicate the potential for genuinely new ways of approaching current problems.

Noise, sound, and music can never be far away in terms of achieving this stated aim. They demonstrate the extent to which any 'bringing to order' – in this case the way music imposes patterned regularity on noise and sound – is a precarious one. As a particular example of compression, music and musical notation can usefully be deployed to support claims for an expanded mode of analysis that can move beyond the deficiencies that are inherent in previous modes and demonstrate the veracity of arguments that foreground dynamic processes of movement and incessant change. Bringing together a great many theoretical discussions around noise, sound, and music will underpin claims that the digital present requires acoustic rather than visual modes of engagement to properly understand its significance.

Victor Zuckerkandl in *Sound and Symbol* (1973) argued that music, rather than other linguistic forms, was best placed to resolve the lingering issues of correlationism. That is the unbreakable chain that not only links, but also separates, the object and the subject, the world as it is and the world as it appears. He adds to the physical and the psychic (object/subject) a third dynamic force evident in music to reveal the process that forms the connection – not a connection between two abstract positions, but one that binds them together as a singularity. It is a dynamic force, a process of constant interaction and communication back and forth, a process that orders and patterns (as a mode of compression) while simultaneously retaining the potential to be otherwise. It is not conducive to the imposition of representational meaning through language (spoken or written) but the possibility remains for its expression in the mathematical mode.

Mathematics, unlike the spoken or written word, is a language with a 'mind of its own' – strictly rule laden, and humans are its adherents. Both types of language (as sound and logic) display different degrees of what is understood here as compression. They combine in music. More about this will follow as sound and music are embedded as features of all art insofar as it constitutes what can be described as a 'relative mode of expansion', embellishing and adorni rather than reducing and defining.

Deploying sound, listening, noise, and music as recursive tropes, what follows will endeavour to plot a sensory journey through compressed and expanded territories of order and chaos such that political arguments can be propagated and sustained. Language, music as a form of language, format (media), and the digital along with the *drive* to compress and the *need* to expand, all raise questions for modes of experiencing, for aesthetics, and for contemporary models of political economy.

Language, in its multiple forms, it will be argued, is a mode of compression. It is 'clear and distinct' in proportion to its variability. Its origin and its philosophical and political implications are explored through the enduring figures of noise and sound showing how compression as a process of reduction can have multiple and sometimes unintended effects. It can be an incredibly useful practice but can equally generate conditions and circumstances that serve to disrupt established orders at best and descend into 'chaos' at worst.[1] At present this fragile balance is insufficiently understood or has not been adequately communicated.

What is important is the outlining of a method. Researching (context), experimenting (with different arrangements of voices), and communicating are comingled practices that in combination form a point in space or a moment in time where/when science, art, and philosophy merge, and where freedom of expression and disciplined processes are equally important. This was the gift that Michel Serres gave the world, and his thinking along with that of Alfred North Whitehead, among others, figure largely here. Recognizing what happens when different voices are arranged to form dialogues within a reconstituted spatio-temporal context requires the interplay and simultaneity of compression and expansion to fully appreciate the complexity at work in the universe and our ability to comprehend it.

This method reflects on the speculative turn throughout. Speculation, like Heidegger's questioning, masquerades as a counterpoint to compression. On one level it as an expansive linguistic practice, but ultimately it seeks to compress in the pursuit of essence, a pursuit that for some (Graham Harman, for instance) is ultimately futile. Whitehead on the other hand skips right past this laborious task and takes as given the existence of a real that is in a constant state of flux, decay, and renewal. In the end then, it is the way(s) in which human beings

[1] Chaos is a term that of course takes numerous forms. While used in the general sense of disorder in this instance, it must always be recognized that chaos retains, like noise, positive properties.

comprehend, and their status in a material universe of objects, entities, events, that is at stake.

Our experience of the world is modal. The totality of the universe, as singular, monistic *and* relative forms the ground or materiality through which we embody different, distinguishable but not separable modes of knowing. To a degree our relationships, with ourselves, each other, and the wider universe are purely physical, biological, and conform to specific cause and effect relationships. In another mode however we are perceptive beings who organize sense data into ordered patterns that are abstracted as symbols. Finally, as advanced organic beings we evolve beyond both cause and effect and direct recognition to be able to contemplate in such a way as to both organize and order data into percepts and concepts while simultaneously recognizing the fundamental ground of our originary physical and biological state. This process, and the ways in which the different modes communicate, is best understood via recourse to sound rather than vision, it will be argued. Combining art, music, sound art, philosophy, and science, the aim is to *get sound into the image and time into space.*

In doing so, a kind of spatio-temporal relativity emerges from the interplay of permanence and transitoriness, provoking a response to ocularcentrism and a recognition of the significance of sound, or more specifically noise (as a means of engaging with the fleeting and only relatively stable). This in turn necessitates a theoretical dimension, a technical dimension, and an aesthetic dimension. Also in the final assessment, one that can engage at the level of political economy based on a *sonic economy* of the senses.[2]

Compression is understood here as a process, not a metaphor, that itself takes many forms and is effective across multiple modes – not least the digital where it is vital in terms of the relationship between the real and the virtual, or more accurately in terms of understanding the virtual as an aspect of the real.

One of the most significant themes to be explored is the technical process of compression as it applies to music, particularly the MP3 algorithm. If understood correctly, musical compression, as a mathematical model, is a mode of simultaneous compression and expansion, isolating elements of an actual entity into its constituent parts and then forming them into a unit. All the while the discarded elements stand ready to be recuperated. In breaking down it opens up. It reveals (or it can do in the hands of an artist), identifies, and displaces the

[2] This term is defined more fully in Chapter 2.

material that is not designated as necessary. And as more than just a convenient metaphor, it is an active and operative process where even the concept itself divides into distinct modes – lossy and lossless.[3] Mathematics as a mode of engagement then, and its relationship to music as a non-representational mode that works through geometry and the algorithm, shows how order and chaos can coexist in shifting ratio without recourse to a universal imposition of value.

Michel Serres and Alfred North Whitehead are the two key figures in what follows, and their considerable contributions will be mixed with ideas from Isabelle Stengers, Deleuze and Guattari, Michel Foucault, Ray Brassier, and Wittgenstein, to form the latest in a long series of interventions that call for a new mode of engagement. The onset of the digital gives this a new dimension, makes it more urgent but also presents us with new opportunities. The interface between actual entities as substance, both physical and conceptual, will be explored in relation to classical and contemporary positions. Rationalism will variously be cited as a system of human understanding and as a process common to all thinking beings, its enaction not particular to any one being, and further as a system itself, an event, a coming together of energy and particles, conductors, connectors, and so on. Not rationalism as such but an emergent facet of interaction and communication. This is the system that Serres describes in the *Origin of Language*, and it follows Whitehead's process, where events as actual entities are not subjects but superjects operating through the flux of energy rather than differentiated Newtonian particles. Process, dynamism, and vibration come together to support monistic positions where subject and object merge to create a universe that can be better understood through sound.

To develop a practice/process of engagement based on sound and listening is also to develop Whitehead's distinction between everyday lived experience (notions) and a more systematic mode of engagement that must include an

[3] Audio codecs are programs that compress data for transmission and decompress that data on the receiving end. Their speed is measured in thousands of bits processed per second, known as 'bitrate' or 'kbps', and this number varies even within a single format. Generally, a lower bitrate means a smaller file, but that also means more data (sound) is lost in compression.

Lossy audio formats lose data in the transmission. They don't decompress back to their original file size, so they end up smaller, and some sound waves are lost. Artists and engineers who send audio files back and forth prefer not to use lossy formats, because the files degrade every time they're exported.

Lossless formats decompress back to their original size, keeping sound quality intact. Audio professionals want all of the original sound waves, so they prefer lossless. These files can be several times larger than MP3s. Lossless bitrates depend on the volume and density of the music, rather than the quality of the audio (https://www.adobe.com/uk/creativecloud/video/discover/best-audio-format.html; accessed January 2024).

acknowledgement of the difficult idea of 'importance' – what's kept and what is discarded as a conscious (or not) process and the ability or inability to control the source of sense data. The dynamic energy specific to sound is a way of negotiating the relationship between these modes (as the experience of everything processed into the comprehensibility of something) and can act as a model for a more general assessment of sensory engagement with the world.

As an addendum to post-structuralism, to the linguistic and speculative turns, immune (hopefully) to the vagaries of fashion in philosophy or any other walk of life, the enduring questions of duality, dialectics, state power, smooth and striated space, universalism, and difference are unashamedly revisited and rearticulated here. Another attempt, this time composed and mediated through sound, to amplify the call to think again, to think differently about ourselves and our circumstances in the face of new problems/challenges, as well as some familiar ones. To do so requires engagement with a multitude of voices, a cast of players orchestrated in the service of a moment, a pause to listen – one that is recorded (in words) ready to be experienced anew in always altered circumstances and environments – destined always to dissipate rather than constituting a claim to permanence or solution, vibrating and vacillating between the local and the universal, the specific and the general, never entirely lacking order but always inclined towards chaos, making sense as all things do, relatively.

Today language as it has been understood for so long is giving way to data. The digital reduces the vast potential of the universe, and experience of it, to units of calculation that can be compressed and ordered to create sets, categories, disciplines and 'meaning'. It creates a kind of super-linguistic framework that more than any of its forebears neglects much that needs to be accounted for (in a sonic economy) in order to grasp and make sense of the vast potential it has helped to unleash. A data driven world though is no less noisy. Like language, it reproduces at an alarming rate, creating information overload and with it the necessity to compress.[4] Like all language the digital generates its own 'waste' material through a process of filtering. An essential element in what follows will be to examine how this works in practice in multiple modes of experiencing. For Michael Oakshott, philosophy as one mode of experiencing, rather than being

[4] Why is the world suddenly characterized by an overwhelming and bewildering volume of information? That it is, is well documented ('information overload paradox'). In some ways however it is nothing new (https://www.economist.com/1843/2018/06/21/information-overload-is-nothing-new).

an encyclopaedic compendium of curiosity driven knowledge acquisitions, is rather a coming together of information to form resonant events, filtering out what is not needed.[5] Likewise, Serres in *Geometry* muses on the excess of information and the extent to which mathematics, as another mode, filters out what went before rather than simply adding to it.

The act of filtering out or removing superfluous information, as well as being a mode of compression, also constitutes a kind of cartography, a symbolic representing of territories and spatialities, as well as a more general mapping of perceived data into comprehensible arrangements. In their digital mode, maps are particularly adept at both compressing and encompassing; they seek to capture everything while simultaneously shrinking what it is they stand in for into a space, or non-space (in the digital realm) without extension or quality. Once in this space they can be compressed further, removing all unnecessary impediments to efficiency.

Digital mapping comprises representing, repeating, 3D modelling, spatial modelling, hyperspectral modelling, and so on, and will be considered in due course in relation to ontology, hauntology, the unseen, dimensionality, materiality and the object, materiality of the past, present and future, encompassing both its spatial and temporal features. It will be important to consider whether such mapping produces a digital twin, or copy, a reflection, echo, double, or a resemblance as representation. As a symbolic mode of representation, mapping is an abstraction, it resides in the linguistic realm where it can be variously ordinary or extraordinary – it is entirely real, an actual entity, an event, existing somewhere between language and mathematics. Digital mapping as a process also relates to noise, sound, and music on a number of levels, not least the need for static representations to be imbued with temporal features.[6] Sound is the last terrain to be mapped, because of its spectral nature – its presence falling away into the past as it simultaneously traverses into the future, but its eventual capture is significant as will be made clear in what follows, as it challenges fixed modes of representation in favour of mobile and dynamic configurations.

The book is divided into three chapters. In the first, *A Sound Argument*, a position is formulated that seeks to establish sound as a significant factor in

[5] Oakeshott, M. (2015), *Experience and Its Modes.*

[6] 'Reflections, echoes, doubles and souls do not belong to the domain of resemblance or equivalence; and it is no more possible to exchange one's soul than it is to substitute real twins for one another.' (Deleuze and Guattari, 1994: 1)

terms of critically assessing the idea that our investigations should be channelled *through* sound and imbued with a sonic sensibility.

In his introduction to Deleuze and Guattari's *A Thousand Plateaus*, Brian Massumi confesses he doesn't quite know what to do with a book where the authors recommend you read it as you would listen to a record. He says:

> When you buy a record there are always cuts that leave you cold. You skip them. You don't approach a record as a closed book that you have to take or leave. Other cuts you may listen to over and over again, they follow you. You find yourself humming them under your breath as you go about your daily business. (2004b, xiv)

This reads perhaps more like a description of the digital music landscape, and is interesting now, and somewhat paradoxically, in an era characterized by a certain nostalgia for vinyl recordings and the presumed 'integrity' of the Long-Player format – now often listened to in the order intended by the artist. How do you listen to a record? Is there a correct way? Is it as free and as fluid as one might hope? Probably not, but that is not really the point. It is more important to raise the fact that too often assumptions are made about music without adequate critical reflection. This book sets out to correct this. It might be read occasionally, intermittently, in an array of possible orders, but actually the intended order has been very much prescribed (as are many records) as an arrangement of voices. When read (almost certainly silently and not aloud) the ideas and their relation to other ideas, whether recorded in the text, or alive in the readers head, are transitory, temporal, and hopefully agile, alert to new contexts and combinatory drives, settling and unnerving in equal measure (but never measurably equal), as a mode of compression that anticipates expansion.

Chapter 1, 'A sound argument', will examine in turn:

1. The process of digitizing music
2. Postmodernism and the search for truth among the noise
3. The speculative turn
4. Improvisation
5. Serres on sound
6. Cox on materiality
7. Boulez – smooth and striated – serialism.

In Chapter 2, 'Language as a mode of compression', a prolonged and detailed explication of the ideas of Alfred North Whitehead and Michel Serres is set out.

Specifically, but not exclusively, Whitehead's *Symbolism* and Serres's 'Origin of Language' are brought into concert to account for systematic processes of interaction and change that in turn underpin the use of compression as a contemporary mode of experience and as a figure capable of accounting for contemporary digital forms of symbolic representation.

The third and final chapter, 'A geometry of listening', combines aspects of precise measurement and pattern formation with dynamic topological regions and flows to embrace the simultaneity of representational and non-representational modes of experiencing. Drawing on Whitehead's point-free geometry it cites noise, sound, music, and listening as mechanisms for understanding complex and interoperable processes.

It also examines in detail Michel Serres' text on geometry and infuses it with a sonic dimension before moving on to similarly account for cartography and digital modelling as modes of compression.

1

A sound argument

Why sound? Because everything vibrates, moves and is forever in process. Where vision fixes and makes permanent, the sonic contemplates transitoriness with the necessary dynamism.

In increments the figures of noise, sound, and music can be deployed to articulate and support a philosophical process that calls for a reconstituted sonic sensibility. In doing so it is important to provide some prior context as to how this might work and why sound is regarded as such a significant element in the formulation of contemporary critical positions.

In the chapters that follow, the relationship between language, as a mode of representation that takes many forms, and compression, as a process that confers a measured degree of order on a largely chaotic universe, is set out in detail. Compression initiates the loss of a sense of wildness, uncertainty, complexity, bewilderment, and chaos. As such it characterizes the epistemological drive to order, conquer, colonize, and conserve that underpins the coming to prominence of the visual. And what is wrong with this we might ask? The need for just such a visual ordering has, it would seem, never been greater. Our contemporary digital milieu and its incessant exponential growth surely requires us to develop suitable filtering mechanisms capable of avoiding terminal information overload. On one level this is a reasonable observation, but it may not be entirely that simple. As we develop such mechanisms, unexpected consequences can occur. While it is assumed that we get smarter over time, it may be that our drive and instinct to 'make sense' of the world in certain ways (predominantly visual) means that with enlightenment also comes a degree of enshadowment or endarkenment. As knowledge is acquired something paradoxically is lost; in the adding to, something is taken away. What that something is can potentially be rearticulated through noise, sound, and music. As such, sonority operates as a more complete model of the universe when it is understood to be a fluctuating constellation that, while sometimes forming predictable patterns, always retains the capacity to transform and evade comprehension.

Images and other sensory data can of course also account for such transitoriness, but when they do, it is argued here, they assume or inherit the temporal characteristics of sound rather than vision. Music is the art that is made out of time and even when, as a mode of sonority, it is at its most ordered and predictable it still retains the ability, in part at least, to generate a sensory experience that can never quite be subjugated to the mode of representation and meaning in the strictest and most reliable sense. Its existence as a melodic organized pattern is always temporally unstable, heard each time in an altered (no matter how slightly) context, forming chains of (relative) predictability and sometimes surprise, linking past, present, and future and always open to resonant and dissonant encounters with other sounds, organized or otherwise.

Recorded and live music clearly differ in this respect, and the history and process of recording sound, whether capturing its dynamism in written form (notation) or via more technologically advanced means, will recur at numerous points in what follows.

In our effort to understand, record, and compress the universe in all its complexity we often succeed only in complicating it even further. Mountains of ideas form in libraries (of Babel), laboratories, and in databases, in letters, numbers, symbols, and notes, that conjoin to form meanings. This ordering in one capacity (mode) or another comes to constitute knowledge as a mode of compression (in symbolic form), yet it also always, inevitably by its very nature, excludes. And when efforts are made to reverse this exclusion bewilderment once again reigns. When everything means something to someone nothing means anything to everyone, and we are left with what might be described as the tyranny of difference, a postmodern deconstructed landscape of options without choices. What is ultimately at stake here then, is the establishment of a framework that supports a mode of experiencing where meaning, in its compressed form, is understood as necessary and functional but not essential – the pattern forms but must be understood as having the potential to deform and reform. Such a (sonic) mode of experiencing must be equipped to process simultaneously, the immanent materiality of interaction between all things and the abstract symbolism as it emerges from the perceptive processes inherent in higher grade organisms (most significantly humans).

To support this aim and serving as a key point of reference for the arguments being developed, it is apposite to consider an artwork produced by Ryan Patrick

Maguire in 2014 entitled *Ghost in the MP3*.[1] In the work, Maguire isolates the data that is removed from an original digital audio version of Suzanne Vega's *Tom's Diner* in the process of producing an MP3 file and reanimates it.[2] In doing so the data discarded by the compression algorithm is demonstrated not to have disappeared but rather to have been set to the margins.[3] The MP3 algorithm enacts a filtering out of sounds that a statistical majority of people are not able to perceive.[4] The definitive work on the history of the MP3 has already been written by Jonathan Sterne. In *MP3: The Meaning of a Format (Sign, Storage, Transmission)*, he gives a full account of its impact, and it is not the intention to repeat those findings here. It is important to say however that Sterne was acutely aware of the wider significance of compression. In an earlier article he writes:

> As a form of data compression, the most compelling part of the mp3 is the psychoacoustic model encoded within it. To personify the technology, it presumes that the sense of hearing discards most of the sound that it encounters, attempting to imitate the process by which the human body discards soundwaves in the process of perception. It preemptively discards data in the soundfile that it anticipates the body will discard later, resulting in a smaller file. (2006: 833)

As such it replicates through science and maths processes that occur both naturally and culturally as will be demonstrated throughout what follows.

What Maguire's work does is to bring to the fore the coexistence of the compressed and the uncompressed; again, not as metaphor but through *dynamic and simultaneous amplification and focus* on a process that occurs everywhere and in everything, including ourselves.

1 https://www.theghostinthemp3.com/theghostinthemp3.html.

2 One of the songs used in the original tests to develop the MP3 algorithm.

3 Maguire says that where the two files are the same or similar, the information in the original audio has been largely preserved in the MP3. However, corresponding time-frequency bins which differ significantly between the two files betray spots where information has been altered or deleted. Different extraction techniques are possible, each leading to slightly different output.

4 Perceptive Coding: There is much more information in a signal than the human ear and mind process. This extra information can be eliminated to make gains in compression. Also, sounds that are perceived, but not as well as other sounds, can be further compressed. A detailed model of human perception must be developed so only unnecessary information is eliminated or harshly compressed. Psychoacoustics, the study of how humans perceive sound, is used to do this.

Some facts about human perception that perceptive coding uses:

- Humans can only perceive sound between 20 Hz and 20 kHz.
- Our sensitivity to sound is different at different frequencies.
- Frequency masking: when a high energy signal is within a certain frequency range of a low energy signal, only the high energy signal will be perceived.
- Temporal masking: when two sounds are very close in time and one is much louder than the other, we only perceive the louder sound (https://www.projectrhea.org/rhea/index.php/MP3_Basics).

The digital turn, and the innovative and creative utilization of its processes, demonstrates how waste can be retrieved and rehabilitated. This happens in a number of ways. In Maguire's work the residual material is reconstituted into something that is distinct from its original context, though obviously not unrelated. The material is reanimated, given a new life, a new-found degree of independence. Once this process is concluded it is extremely important to think how this newly independent entity (or event) might form new articulations with the original configuration from which it was spawned, as well as other entities that it might encounter. If the two now 'new' compositions are played simultaneously, would they from a natural harmony? Not necessarily. It is more likely that the original entity will form a dissonant relationship with its progeny. The discarded material does not stand ready to be reintegrated with the original, as it does with some other compression algorithms, but is designated as residual and surplus to requirement. As both process and allegory then, it stands ready to remind us of the dangers inherent in casting out that which 'belongs'.

The digitization of sound and music has many other implications and impacts. In *Ways of Hearing* (2019), a counterpoint to John Berger's seminal *Ways of Seeing*, Damon Krukowski set out some of those implications that are significant here. In relation to time and our experience of it, there is a distinction to be made between the analogue and the digital, the time of humans and the time of machines. Analogue time is variable, speeding up and slowing down as we experience it. Digital time however is more ordered, regular, and metronomic. Yet it too is faced with a degree of uncertainty. Latency, the split-second discrepancy between sound and its being rendered/translated into code, threatens always to disrupt this drive for absolute consistency. Even in the most machinic of environments then, the potential for chaos lurks meaning organic, mechanical, and digital systems are distinct but not unrelated in this regard. They constitute entangled modes and are relative in terms of perception, processing, and the rendering of data into symbols.

Much of Krukowski's book, and accompanying podcast, centres on his call for us as humans to acknowledge this entangled modal process and to reengage with our sonic environment; to avert our gaze from ubiquitous screens and liberate ourselves from the privatized realm of headphones. He counsels us to stop avoiding eye contact, but even more importantly ear contact.[5] The devices

[5] Individuation and privatization of auditory space constitutes a mode of compression, a retreat from the cacophony of modern life into private space with private means of listening, and ultimately the commercialization of audition where the experience of life is exchanged for specific products substitutes.

used to facilitate this privatized sensory zone and the compression that dictates the sound we hear over them as we listen not just to music but voice is, he says, to the detriment of non-verbal and ambient sound (Krukowski: 62).[6] There is a reduction in terms of what is necessary to convey information and meaning (not unlike Wittgenstein, and entirely contrary to Kittler, as will be discussed later). The signal is accentuated, and the noise (almost) cancelled. This is a double act of compression: the compression/filtering out of ambient noise, and the digital compression enacted by the device. As a result of this double process the information directly perceived loses something. What that is is all the data and material that adds nuance, flavour, style. It may not always be necessary, but its loss should be measured with caution.

Krukowski cites Gary Tomlinson as having a 'beautiful phrase' to express this. Tomlison states that there are 'musical absences at the heart of language'. We are musical beings, Gary believes, as much as linguistic ones. (Krukowski: 69). The assumption here is that music is rich and full whereas language, particularly in its contemporary mode, tends towards brevity and efficiency, and while this may or may not be the case, it certainly merits further consideration.

If 'musical absences' describes that which is present in communication and does not necessarily constitute meaning, then it can be understood as a kind of noise. Its absence, or its confinement to an age before digital listening devices, does not, or should not, invoke romantic regret for a bygone age but a recognition of the constant interplay between expansion and compression as the organic and technical systems get to work. *Ways of Hearing* is in some ways a call to noise, with its richness and complexity, its difference and tonal depth. It highlights, much like Maguire does, the tendency of digital sound technologies to eradicate noise, producing and registering only the signal. 'The real difference is between a world enriched by noise, and a world that strives toward signal only' (Krukowski: 133). For Krukowski, this constitutes loss, not just through

[6] In what follows we will hear a lot from Michel Serres. To adequately contextualize his contribution the following is important:

> Better than it passing and flowing, music percolates. Always so wise in its archaism but scientific enough to rejoin our advanced technologies, our language calls Harlequin's cloak mosaic, a cloak of pieces, designs, forms and colors, diverse and separated, but whose hues and motifs can step across or not the boundaries between pieces, spread their invasion across them, in such a way that this art, at the same time digital and continuous, expands or stretches without tearing and is cut up in the sense of fracture: mixture is also precisely said as cutting. The term 'mosaic' repeats, almost homophonically, the word 'music', its root. (*Geometry*: xxxvii)

the lossy compression that enables it, but also a cultural human loss, and while I take no issue with that, the situation is read a little differently here. The shift to a digital paradigm, even if it does initiate a degree of loss, more importantly makes manifest processes of compression and expansion that are features of all systematic exchanges and all pattern and rule formations, human and otherwise. It is these exchanges that are under consideration here.

It is not the case then, that the analogue and the digital constitute a binary, but that the digital seeks to amplify the signal within a field of analogue noise as part of an ongoing process of compression. Krukowski does himself a disservice when he says he understands the world in the way he does because he is a musician attuned to sound and noise. It is more than that, it is because the universe itself functions in a way that we can recognize as being constituted of noise, sound, and music (in all its diverse forms). This opens up the possibility of richness and diversity that at varying times form patterns and rules that are ordered and seemingly predictable. Such patterns, often constituted as symbolic language, however, do not have to be interpreted in the negative sense but may be conducive to positive and creative interventions so long as they are understood as events, and as transitory energy formations rather than terminal endpoints.

The following chapters in this book show how language becomes detached from its roots, in part due to the will to compress as it turns away from noise, sound, and music. Following a philosophical trajectory that ends with a system of grammatical rules and structures it serves to render the world coherent and ultimately sensible, but in doing so it also makes it static and visual, comprehended predominantly through the eye. Challenges to this stasis in the form of postmodernism/structuralism through the linguistic turn, while restoring an element of indeterminacy and providing more space for interpretation and negotiation, ultimately fail insofar as they undermine any genuine sense of an objective reality, so much so that they are understood by some to be anti-real. This must be revisited. Vacillating between the spoken and the written, the aural and the visual, language – as a bridge between realms of subjectivity and objectivity, idealism and materialism, signifier and signified – leaves much that still needs further discussion if not resolution. Such discussions must recognize the need for a reconstituted sonic sensibility.

More recently, and in response to the perceived failings or inability of continental philosophy during what came to be known as the linguistic turn, speculative philosophy has set out to think anew and initiate some of the discussions required. As an addendum, or perhaps a corrective to phenomenology, structuralism, post-structuralism, deconstruction, and postmodernism, and the 'anti-realism' they embody, the following has been stated:

> 'Without deriding the significant contributions of these philosophies, something is clearly amiss in these trends. In the face of looming ecological catastrophe, and the increasing infiltration of technology into the everyday world (including our own bodies), it is not clear the anti-realist position is equipped to face up to these developments. The danger is that the dominant anti-realist strain of continental philosophy has not only reached a point of decreasing returns, but that it now actively limits the capacities of philosophy in our time.' (*The Speculative Turn*: 3)

The speculative turn denotes a loose conglomeration of thinkers and thoughts that attempt to transcend those limits and capacities. Pausing to consider the significance of some of the ideas generated by this group will serve to strengthen some of the positions being proposed here.

Expressing his own assessment of where we have come to in terms of the postmodern project, Alain Badiou has stated:

> Take one symptom among many: artists, the 'creative' people of our day – choreographers, painters, video-makers – track the self-evidence of bodies, of the desiring and machinic life of bodies, of their intimacy, their nudity, their entwinings and ordeals. They all adapt the inhibited, quartered and soiled body to the domain of fantasies and dreams. All, in the end, impose upon the sphere of the visible the partition of bodies shot through with the noise of the universe. Aesthetic theory merely follows in their wake. (Badiou, 2005: 20)

This describes spatially individuated bodies that respond as desiring machines to chaotic and noisy cosmic conditions that conspire to bring them into being: each one different, and more importantly entitled to be different, every-body to be tolerated, with that tolerance to be enshrined in law de facto and de jure, demarcating random actants floating in a material democracy of objects. For Badiou, the postmodern with its axiom 'there are only bodies and languages' constitutes what he calls democratic materialism, and the equity of difference and the universal valorization of tolerance. However, it has failed us he believes. In its place he proposes a *materialist dialectic*. This is a model of engagement that declares the notion of truth to be both real and attainable.[7]

[7] The idea of struggle is raised by Isabelle Stengers in her contributing chapter to *The Speculative Turn* (2011) in which she critiques Badiou's position in relation to democratic materialism.

To this end he proposes a union:

> If the combined effect of the two French traditions – of Brunschvicg (mathematizing idealism) and Bergson (vitalist mysticism) respectively: the one passing through Cavaillès, Lautman, Desanti, Althusser, Lacan and Badiou; the other through Canguilhem, Foucault, Simondon and Deleuze – is enough to allow this new century not to be devastated by modesty, then philosophy will not have been useless. To produce, in the world such as it is, new forms that might shelter the pride of the inhuman – this is what legitimates us. It matters therefore that by 'materialist dialectic' we understand the deployment of a critique of all critique. (Badiou, 2005: 24)

In supporting this ambition, and paradoxically by deploying one of Badiou's criticisms of postmodern 'modesty' as a celebration of its inability to engender truth, what follows here embraces the idea of 'bodies shot through with the noise of the universe' and rethinks them in terms of creative practice as a means of revealing truth. Noise is not equal to indeterminism, but is the ground from which all bodies emerge, if only temporarily.

In the years since the speculative turn was initiated, it would seem that the urgency for new thinking has been amplified. So, the need should be stated again. This time with a reordered sensory emphasis. The ordering of the senses has led over the preceding millennia to a mode of ocularcentrism that no longer suffices in terms of being able to account for the status of humans in a universe where communication channels between subject and object face renewed ontological challenges. Recent debates around speculative realism, object oriented ontology and a revisiting of questions relating to process philosophy mean that it is necessary to think again, look again, and listen again, to re-engage the sensorial armoury in the service of knowledge. At the outset stands the dualism between transitoriness and permanence, the fluid and the fixed, process and teleology. By rearticulating the sensorium in response to some of these debates a different mode of engagement can be articulated. In this case the orientation places sound at the centre of a material constellation of forces and energy flows where systems form/deform relative to one another, shifting constantly at a vastly different scale and pace.

The intention is not to valorize one sense over the others but rather to bring to the fore some of the characteristics of sonority that compel us to reposition ourselves in a universe of vectors where movement and transitoriness prevail, and where even fixed points in space and 'permanent' structures are subject to the relative vagaries of time. The visual itself, as well as serving to make certain other things invisible, is always prone to an inner vibration, rendered for the most part inaudible but never fully petrified.[8]

The dominance of the visual then always involves a kind of silencing as part of what Jaques Ranciere has called 'the distribution of the sensible'. When something comes into perspectival view something else fades out (at the edges) – seeing is selective, both physiologically and sensorially, but also politically and culturally. Noise, sound, and (sometimes) music on the other hand can serve as reminders of the fact that elements not in plain sight are still present, acting (potentially) as an irritant, a gnawing, throbbing, pulsating presence that does not necessarily seek to undermine the visual (and its ordering) but if understood properly, infects it with its temporal energy, prompting us to look away, look further and deeper, using eyes like ears, alert and inquisitive. Thinking about sound, thinking through sound, and sounding thought, invokes all the senses: merging, clashing, silencing, ordering, losing the self in a cacophony of Dionysian abandon and emerging into a mode of Apollonian clarity where separation prevails.[9] From this point of clarity much has been said, written, and believed about noise, sound, and music. As counterpoints to the visual however, noise, sound, and music are often tainted by assumptions that must be guarded against. The inclination towards uncertainty implies mysticism, and while this can prove fruitful, it can, if one is not careful, fall back into the postmodern abyss.

It has been said that:

> The Greek *theory*, of which we are still bearers and users, is connected with looking, with distance, with observation, grasping, measuring, determining. It is associated with the ocularcentrism of Western culture, with its emphasis on visuality, which implies a supervising, technocratic, and power-oriented attitude to the world. Thinking along with sound, through sonority, through or perhaps from within music and sound is associated with a different approach. With an attitude that favors listening, which is receptive and not controlling, for which meeting its subject has the nature of interaction and permeation, which does not understand sound as an object, but as an aspect of the environment in which we are immersed, which passes through us and thus shapes the very conditions of our presence in the world. (Pokorný and Nitsche, 2021)

[8] To this end sound is always lurking (as evidenced through Eulerian Video Magnification).

[9] See Nietzsche's *The Birth of Tragedy* (1993).

How true is all of this? The description assigns an almost natural ethics to sound that is denied to the visual. While Western ocularcentrism is being critiqued here, it is not deemed sufficient to simply replace it with sonority. Sound too can be authoritarian, determining, and controlling, and equally subject to dominant narratives and discourses that over time become enshrined as normative value laden concepts, as well as imperial and colonial signifiers and drivers. The aim here is not to malign the visual but to embed the sonorous and related temporal dimensionality into a critical framework that properly recognizes noise, sound, and music as factors that operate as more than abstract emotional responses to an eternally mysterious and inaccessible realm. Rather, noise, sound, and music are understood, or will come to be understood, within the framework of process philosophy that challenges the strict subject/object dualism that is an essential element of ocularcentrism.

In embedding a temporal, sonorous dimension into the process of sensory engagement intended to actualize a thinking through sound, it is important to recognize the presence of rules, grammar, and algorithms that serve to modulate excess and which despite acting as limit, guide, and regulator in terms of creative agency (for Badiou the 'creative people of our day') also offer the opportunity, if one chooses (if indeed there is such a thing as choice) to take it, to ignore them and do something unpredictable. Ray Brassier has discussed this in terms of improvisation, and in doing so helps us to understand bodies (artists) as 'shot through with the noise of the universe'.

In his text written for a performance with Mattin at Arika's festival episode 4 *Freedom Is a Constant Struggle* on 21 April 2013, Tramway, Glasgow, Brassier asks, 'What are the conditions for free improvisation?'[10] To address this question, he says we must first state the difference between the concepts 'free' and 'improvisation'.

At the root of his argument is the significance of pattern formation both as biological impulse and social precedent. The free-thinking subject is effectively declared moribund in the face of interrelated sets of objective pattern formation. To convince us of this Brassier distinguishes between freedom and voluntarism. A voluntarist conception of freedom exists as a phenomenon that appears as if from nowhere and unhindered by any cause. He regards this as 'objectionably metaphysical'. In its place he proposes the idea of an act acting upon itself where

10 http://www.mattin.org/essays/unfree_improvisation-compulsive_freedom.html (accessed 7 July 2024).

that act and not the self is sovereign. So, the self is reduced to a set of biological and social impulses that constitute an act, and freedom and the extent to which choice and improvisation take place at all are contingent upon an adherence to patterns and the recognition of the difference between *for a reason and because of a reason*. In the final assessment it is the interplay of these objective conditions that creates the requisite conditions for the self. Acts *for a reason* form patterns than in turn generate rules that then come to frame acts *because of a reason*.

Brassier also distinguishes between patterns (biological) and rules (cultural), going on to say

> The ability to act is composed out of two distinct strata of behavior: that of pattern-governed behavior on one level, and that of rule-conforming behavior on the other. The act results from the superimposition of these two levels, i.e., from the superimposition of rule-conforming behavior onto pattern-governed behavior. It is the product of the intrication of these two levels, but it cannot be reduced to either. (Brassier, 2013)

Brassier chooses to discuss this in terms of noise, music, and improvisation, beginning with the dance of the bee. The 'dance' does not pre-exist but is an outcome of a pattern that is itself not predetermined – it is instinctive and enters the domain of rules only as an outcome of the initially unintended pattern. *For a reason* becomes *because of a reason*, pattern becomes rule – the notion of an act that precipitates the reflexive ability of acts to operate in often unexpected way is a process that can account for the creative act without recourse to an abstract self (ex-nihilo). Stepping into the realm of noise to make his point Brassier situates himself along a temporal continuum where pattern formation, recognition through the engagement of an eventual self, and the generation of the social from biological impulses emerge from noise as the preconscious (where consciousness is an act on an act) ground for what will conspire to become music just as biology becomes culture – different but inseparable.

He says:

> The ideal of 'free improvisation' is paradoxical: in order for improvisation to be free in the requisite sense, it must be a self-determining act, but this requires the involution of a series of mechanisms. It is this involutive process that is the agent of the act – one that is not necessarily human. It should not be confused for the improviser's self, which is rather the greatest obstacle to the emergence of the act. The improviser must be prepared to act as an agent – in the sense in which one acts as a covert operative – on behalf of whatever mechanisms

> are capable of effecting the acceleration or confrontation required for releasing the act. The latter arises at the point of intrication between rules and patterns, reasons and causes. It is the key that unlocks the mystery of how objectivity generates subjectivity. The subject as agent of the act is the point of involution at which objectivity determines its own determination: agency is a second-order process whereby neurobiological or socioeconomic determinants (for example) generate their own determination. In this sense, recognizing the un-freedom of voluntary activity is the gateway to compulsive freedom. (Brassier, 2013)

Free improvisation then is the manifestation of a process, in this case the involutive process. (In Whitehead this will be named perception in the mode of causal efficacy, but more about that later.) It is the act acting on the act and is a kind of hyper-example of processes that inhere naturally and culturally, becoming amplified momentarily and, in some cases, for just long enough to enable our understanding (if that is indeed the correct term, as the act of understanding is itself an act) of them.

Such a model of free improvisation doubles as a 'musical' description of a dynamic system of pattern formation that has implications beyond sound. It constitutes a sonic sensibility that can be expanded to explain socio-political and economic factors: patterns and rules now constituting a truth free from agency that has in turn been relegated to the status of 'second-order process'.

In an exchange that is prefaced on what can be understood as 'musical' improvisation, David Roden (2019) challenges aspects of Brassier's position. Both use their thinking to explore notions of agency, freedom, and structures, that support modes of engagement beyond improvisation itself, ultimately generating new kinds of ethics and politics. Roden suggests that Brassier's text, for all it is rooted in musical improvisation, never really shows how the process of musical improvisation works in itself. If for Brassier the emergence of rules is the manifestation of a process that is in operation all around us and in us all the time, for Roden it is specifically a musical mode which, when explored in detail, serves to show that actually there are no such rules at play at all.

This embodies the idea that there is something non-linguistic about improvised music as a form of expression that is honed to generate affect rather than meaning and it is therefore of a different order to formal linguistic rules and patterns. Roden says: 'According to Brassier, this structure is reflexive. It requires, first of all, a system that acts in conformity to rules but is capable of representing and modifying these rules with implications for its future behaviour' (Roden, 2018).

At issue is the existence of rules/codes that act as triggers, and the ability to operate contrary to them and to adopt other coded determinants/options and whether this constitutes a kind of 'self-determination within a causal order'. For Roden, Brassier's position retains traces of the subject and falls back on a kind of neo-rational Kantianism insofar as it uses the ground of natural and biological pattern formation to support the emergence of the figure of a rule-dependent subject. Roden reads Brassier as having enacted or re-enacted a bifurcation.

I would say on the contrary that he enacts a conjoining, with *pattern* and *rule* constituting specific modes of a singular energy-based reality. Such an ontological flattening should not and does not go so far as to eradicate the human subject, but rather reorientate their position within a material plenum that can account for subject and object without existential distinction.[11]

As Brassier uses the act of 'musical' improvisation to account for his position, so Roden uses music, and specifically jazz improvisation, to counter that position. On the one hand it is said that Brassier fails to demonstrate clearly the relationship between patterns and rules. On the other hand, for Roden, this is because the existence and validity of the rules themselves cannot adequately be demonstrated. What can be demonstrated conversely, he believes, is the absence of rules, leaving only a freer, more random ground of pattern formation described as speculative posthumanism.

> Brassier's position arguably depends, then, on the conception of an interpreting subject it is not in a position to satisfactorily explain. Despite its pretensions to naturalistic virtue, his universe bifurcates between a non-normative material reality and an order of thought that depends on it without really belonging to it ... These metaphysical issues lurk in the background in Brassier's short text on improvisation – particularly in the claim that the act of improvisation involves an encounter between rule governed reason and pattern governed mechanisms. Brassier does not specify how such rules operate in music, or how the encounter between rules and mechanism can occur.[12]

11 Brassier reflects on the work of Bruno Latour and the saturating effect of the postmodern saying,

> unless reason itself carries out the de-mystification of rationality, irrationalism triumphs by adopting the mantle of scepticism that allows it to denounce reason as a kind of faith. The result is the post-modern scenario, in which the rationalist imperative to explain phenomena by penetrating to the reality beyond appearances is diagnosed as the symptom of an implicitly theological metaphysical reductionism.

Such a 'reductionism', should that be the correct description, might also be understood as a mode of compression as a de-mystified aspect of rationalism. Brassier identifies Latour as a key exponent of the anti-rationalist position (Bryant, Srnicek and Harman, *The Speculative Turn*, 2011: 51).

12 https://www.performancephilosophy.org/journal/article/view/230/347 (accessed 7 July 2024).

In order to adequately position the debate within the context of the argument being developed here, a summary is useful: Brassier develops an argument that accounts for a restaging of the self and subjectivity within a universe of patterns and rules, one which draws on the act of free improvisation – free from the self. For him the process of pattern formation provides the ground from which emerges recognition of those patterns and their function in the form of rules that evolve into social convention. The act of recognition requires the combining of pattern and rule and eventually this combination operates through what we come to understand as consciousness or self. But it is in no way an abstract self. On the contrary, Brassier's explanation understands the act of selfhood as immanent, describing a system of patterns, adherence to rules, but also the reflexive ability (of the system) to enact responses that operate contrary to the rules, but always as a pattern in their own right. This position accounts for the existence and importance of a *kind of subject*, understood as both pattern *and* enactor of patterns, without recourse to a self that is 'objectionably metaphysical insofar as it invokes entities and forces that are dubitable, at the very least'. What is being objected to is a mode of spiritual idealism that separates the self/subject from the object. Instead,

> pattern-governed behavior is ubiquitous in the biological and physical realms. Physical systems realize complex patterns without intending them. The pattern is incarnated by the components of the system, each part of which constitutes it, but the constitution is effectuated by something as mindless as a wiring-diagram. The latter mechanism codes for the pattern, without the structure of the pattern having to be represented by any part of it. (Brassier, 2013)

Recognition of such non-representational patterning allows us to deploy an acoustic or sonic mode of engagement where sound resolves to pattern as music without specific meaning or correlate. It can accommodate normalized rule bound modes at the same time as being able to transgress those patterns that have mutated into rules/norms. This speaks to a certain kind of chaotic system, one where order teeters on the brink, and is well described using sound and music as a guide (or wiring diagram). The systematic patterning is dynamic and live, it is the moment that constitutes improvisational time, and it cannot be captured or rendered into formal language. To attempt to do so would expose it to rules, coded into a new kind of pattern that is itself inherently unstable.

The point is not to eradicate rules altogether as Roden does, but to recognize and navigate their contingency. Genuine agency involves the capacity to follow or deviate from a rule. What we understand as agency might be better described as the capacity of a system to tolerate and encompass unexpected responses, interventions, and challenges – sometimes the unexpected is accepted and integrated, constituting what might be thought of as an enhancement or a development. At other times the unexpected is treated as being problematic or disruptive. In such cases it is filtered out, compressed, and marginalized.

Roden believes that the lack of a satisfactory explanation of the way in which patterns and rules interact means that the abstract subject retains a privileged status in Brassier's account. As an alternative he presents a different kind of chaotic system that, despite being elucidated through music and specific modes of improvisation, dispenses with the idea of rules and the subject altogether, and is left only with patterns that presumably enter into free-form relationships. In the absence of any kind of subject, free or otherwise though, it is difficult to extend the model much beyond the free improvisation that is being described. Both Roden and Brassier attempt to generate a 'musical' model that is capable of extending into other realms. However, the absence of rules that Roden identifies in his model, replaced by affective response, though it is extremely interesting in a strictly musical context, does not seem to be applicable in realms where selfhood remains significant, and is best understood in the way Brassier explains.

The difference in positions seems not to be much of a difference at all but rather the enactment and emergence of modal patterns that in this case are slightly dissonant but with both serving to score a process, making audible that which for the most part remains invisible. There is only the system (as opposed to bodies and languages), always patterns and rules that exist in some mode or other, a configuration/reconfiguration of a set of possibilities. The ability to reconfigure the 'rules' is an operation that is perpetuated by what Michel Serres called 'the guardians of the possible'. In his book *Rome* he describes termites operating in a mode that is akin to free improvisation, placing balls of earth some place other than where they have been prescribed and building in the process complex and irregular structures. Brassier presents his argument from a similar perspective, but in relation to 'freedom' and improvisation in 'music'.[13]

The interplay between Brassier and Roden, then, brings to the forefront important ideas around rules, patterns, repetition, and difference that play out over time. This 'playing out' is important, and though it emerges from a

[13] Based on a reading of Wilfrid Sellars 'Some Reflections on Language Games', *Philosophy of Science*, vol. 21, no. 3 (July 1954): 204–28.

discussion around music and improvisation specifically in this case, it speaks to much wider concerns. If an incessant relativism pervades the universe as it beats in and out of time, slowing, pausing, to form recognizable (audible, sometimes visible, but often not) patterns while simultaneously readying itself to move on at varying speed, then any sense of certainty must necessarily be tempered with an assured anxiety for what might come after. The humming of the motor of change, though it can be momentarily quietened via recourse to a measured and primarily visual expression of predictability, cannot ever be truly silenced. It is this pervasive presence that underlies the fact that whenever chaotic systems resolve to some kind of order, they do so through a process of compression, a process that often excludes that which does not fit, that which irritates, or most importantly that which undermines any claim to authority the established order might make. This is true of all systems – organic, mechanical, thermodynamic, quantum, social, and political.

While admittedly Brassier doesn't go into any real detail on the relationship between rules and patterns as rudimentary configurations in his description of unfree improvisation, others have provided more than satisfactory accounts that evidently tally with his thinking. It is to those ideas that we now move, to consider how rules and freedom from rules relate to smooth and striated space-time. This will be addressed shortly in relation to Pierre Boulez, and in more detail in subsequent sections in relation to Deleuze and Guattari's deployment of Boulez's categories in *A Thousand Plateaus*. But first it is to Michel Serres and his deployment of the figure of noise that we turn.

Embedding noise, sound, and music in visual art, and at the heart of knowledge generation, as well as other forms of production (economics), is central to the task at hand here. Noise is deployed by Michel Serres to resolve some very complex scientific and linguistic problems, and the entire subject/object dualism collapses in its wake. It also establishes a set of statements and theoretical manoeuvrings that can serve to address many of the very significant issues that will be explored more fully later.

For Serres noise constitutes a kind of fundamental ground. It is everywhere and in everything, not merely an obstruction for information and meaning but its constituent and essential ingredient, its medium. If our experience of the world is reduced to rational meaning, certainty, measurable and accountable phenomena that can be made sense of, relied upon, and exchanged in terms of a linguistic as well as a material economy, then certain difficult issues arise. As processes of compression such rational systems create residue, waste, as 'silenced' detritus. But silencing noise does not eradicate it entirely, as Maguire has shown. It exists beyond its comprehension – lingering, stirring, swirling, exorcised from its systemic context, diffused and dissipated, but always ready to join a new system or to undermine the original source.

Some examples of the deployment of noise, sound, music, and hearing form Michel Serres's oeuvre:

In *Genesis* 'hearing is a model of understanding. It is still active and deep when our gaze has gone hazy or gone to sleep. It is continuous while the other senses are intermittent. I hear and I understand, blindly, when evidence has vanished, and intuition has faded out: they're the exceptions' (1995: 4). Evidence and intuition tend towards exclusion whereas sound proffers towards fullness, and this is key in terms of the proposition that compression is a useful term when attempting to think through issues relating to perception and reality.

In *The Five Senses* 'noise means relation, passage, variation, invention, for it arises in the spaces between fixed points and positions. But it also means excess, chaos. Noise is both a matrix of possibility and the cauldron of indifference in which true invention is ground down or swallowed up' (Connor, in Serres, 2008: 10). From the matrix of possibility patterns emerge to from resonant events as actual entities that are selective in their formation, but which always remain susceptible to the indifference of the whole. Just as compression and expansion corelate so do noise and its progenies, and they do so by means of the transmission of messages and codes. These messages and codes take many forms.

In *Hermes* we encounter cacography:

> Modern knowledge expresses itself and is understood in terms of codes – whether in the domain of the unconscious considered to be structured like a language (pp. 80–1), or in the domain of the life sciences, which teach us that the transmission of life is a function of the genetic code (pp. 72–9), or in the domain of the experimental sciences such as physics and chemistry in which codes are essential. Modern science is thus specifically concerned with the study of all aspects of the transmission and propagation of messages – information, noise, redundancy. (Literary criticism understands these same problems in terms of theories of code, language, writing, and translation.) In this respect, information theory is all the more relevant, since it is located at the crossroads of all fields of contemporary knowledge. (1982a: xxiii)

The process of coding is central here, as underscored by the work of Ryan Maguire discussed previously, and it will recur as we proceed. If coding strives to achieve *order* (as noun) we must emphasize *ordering* (as verb) to maintain the vital potentiality of even that which has been discarded; especially that which has been discarded, for it retains potential whereas that which has been ordered and successfully prescribed is spent.

Hermes also contains a chapter 'The Origin of Language' that will be discussed in considerably greater detail later.

In *Parasite* we encounter cacophony where noise is constituted as the condition of possibility for the universe and its elements to harmonize, to come together to form resonances however fleeting and improbably rare. This rarity is hope whereas certainty and repetition are death. The conquering of noise bequeaths stasis. This potential is now given the word 'accord' as a mode of becoming harmonious:

> The accord. On what object, about what? Later, later. But, at least, on a meaning. Later, I tell you. On the sound then, first of all. If you do not make too much noise, I shall try to stop my own; if I sound correctly enough, you will evolve toward that peaceful justice. Before exchanging a single word, before agreeing on the code, we must at least emit a sound together. Here, one can send and receive at the same time. Yes, my signal is the only one in the world, and my voice cries out in the wilderness, in the stony desert of my whinings. Specific, individual pebbles. Let us remove these parasitic stones; by filing down the thorny edges of the sound, we approach each other. Being in tune musically and sonorously, the accord is the archaic accord of nuptial agreements [accordailles]. Together. A vibration in several voices. Coming together [jouissance]. The collective, at the least, is sonorous utopia. Hermes requires translations. Pentecost sings, resounds, and blows; the tongues melt in this fire; music has spoken in tongues. It is free of parasites. The universal language of a buried contract. (1982c: 134)

The idea of 'filing down the thorny edges' of sound is what creates music, emerging as an accord, a moment of confluence that needs no translation, a compression of all the possible connections, all possibilities, the sets or subsets of lonely monadic elements combining to constitute event, or accord, while all around them chaos reigns. And not just music but all events, all confluences, all subject, all objects – everything emerges from noise.

Also in *Parasite*, Serres deals explicitly with noise and music, and it is worthwhile dwelling on these accounts to support the here-stated intention to answer the question of why sound is so sensorially significant?

On noise: The universe, as Leibniz concludes, consists of independent entities (monads), each generating noise of their own – not speaking, not listening, combining only to generate noise. Yet miraculously, on occasion, they find each other in a more harmonious register.

He said:

> here is an orchestra. Each musician plays his instrument as if he were alone in the world. He likes only his English horn; this English horn is he himself in person. He plays his part of the score, and when he has finished, at the very end of the page, he puts down his things and leaves the theatre. But only to die. How could the first viola be in harmony with him, for the first viola has never thought of any thing but his four strings?

Leibniz answered:

> God created the viola so that at a certain predetermined and well-fixed time, it would produce the note preformed harmonically to that of the English horn. God foresees harmony and God is harmony. History is programmed; everyone has a score. Others say that they are in the same linguistic milieu together. Words have to find each other, since they are part of the same set. And this is the same solution: there is a conductor or a common text to play. Someone or something always precedes. (1980: 121)

This improbability becomes everything to us. From the vastness of the universe entities entwine to form resonant events, including humans, as both objects and 'subjects': 'Fortunately, the rare exists, exceptions come about, novelty appears – the improbable miracle. Through this rarity, the world comes into existence, we live, and we think. These three events are improbable but are there nevertheless' (1980: 122).

To further develop his thoughts on music Serres draws on some lesser-known facets of Jean-Jaques Rousseau, who worked as a composer and transcriber of music, and used this experience to sound out his argument.

> Just as a hungry guest is fascinated by the buffet of a feast, Jean Jacques keeps (to) music. There lies, manifest and under wraps, the solution. This solution includes and excludes him; it includes him as a partition, and it excludes him as a parasite. Who am I? The condition of music, of the accord, and its obstacle. He copies it, writes it, keeps it, watches it, gives it. He exchanges it and sells it. He has always known how to resolve his question: not by what he says, but by what he does, which is clearer than the light of concepts and easier to understand than the doubletalk [langue de bois] of philosophy. (1980: 134)

Clearer than light and easier to understand than philosophy, music is the sound of accord. Its very presence and emergence from noise is testament to a chaotic universe that is driven towards order, but which equally faces an existential threat from its own ordering. The individuated self as partition (as Badiou reminded us) emerges as musical accord, and almost immediately begins to inhibit its majesty. Explained via recourse to a God-person-pre-text (score/instructions/algorithm) this musical accord is so profound as to compel us to try and enshrine it, fix it, and measure it. In the text/score sound does not precede but follows the improbable harmony and where it reigns it spells death and stasis. There is nowhere left for it to go. But even when scored and recorded, music strains at the edges, as an energy that cannot be forever contained. More than a picture in a frame it is restless and prone to attack from parasitic noise and new encounters.

What is it that brings everything together? Score, partition, conductor? These questions draw us back to Brassier and Roden and to the ways in which patterns and rules interact. For both, sound and music, and modes that embody them, present opportunities to untangle this interaction. Noise, sound, and music should then be deployed as modes of engagement with a transitory universe vacillating between order and chaos, negentropy and entropy, where despite attempts through notation and recording to stabilize its nature, its inability to work reliably as a mode of representation turns out to be its biggest gift.

This mode of engagement, infused as it is with sonic sensibility, is not, as one may suspect, an absolute. Like smooth and striated space or the body without organs (to be discussed in due course), representation and non-representation, compression, and expansion, it is a relational mode wherein aspects of chaos and order, noise and music are in concert, with ratios shifting constantly; sometimes, hesitating, coming to rest (or almost rest) before moving again.

This combinatory tendency as pattern and rule in music is important – interpretations, movements, version, remixes, cut/copy formats, and so on. Without diverting too radically from the stated path, we can identify in tonal music and in serialism, and latterly in hip hop and electronic remix culture, evidence of music's willingness to be recontextualized, reborn, returned in new formations as pattern and rule collide. This is why sound serves as a useful model. It takes many forms simultaneously without ever quite conforming to a universal set of criteria or rules. It is as happy as pattern *or* rule, or as noise *or* music – scored, improvised, ordered or wildly and extravagantly free, amenable to code, language, spatial and temporal measurement, and universally present, even in the other senses, as vibration, as energy that is the ground of possibility without author or pretext. It is virtual, real, and actual.

If beings themselves are vibrating energy fields, we should consider what it means to be in tune with the world. The answer is, not to be taken too literally, or encumbered with mystical ecological discourse, but simply to be alert to the presence of other factors, conditions, contexts that morph and shift towards ultimate breakdown and recombination. Thinking through noise, sound and music opens up new forms of knowledge across all the sensory realms and modes of experience. Importantly, in the context of critiquing ocularcentrism, the visual does not escape entirely. As well as cacophony there is cacography, and this will be considered in more detail presently in relation to the thinking of A. N. Whitehead and the creative potential of error. The visual, however, has historically been more adept at fixing its gaze, compressing its object, and banishing excess to the shadows.

With this in mind, another key contribution to the debate around the significance of sound should be considered. In his article 'Beyond Representation and Signification: Toward a Sonic Materialism' (2011), Christoph Cox argues that sound has effectively been eliminated from the artistic process. Art has been compressed, and sound set to the margins. This is a problem. Not only does it mean the dominance of the visual over the audible, but it also strips the visual of an important temporal dimension. An examination of his position supports the argument that the introduction of noise and sound into all artistic practices and indeed to all aspects of life serves to counter the limiting effects of this kind of compression on aesthetic as well as political and economic grounds. How? The point is not to argue for a world of chaos, or to enact a return to the prehistoric, but to forge new kinds of knowledge and practices that are able to accommodate uncertainty, transitoriness, and the basic principle that to order is often to exclude, and the constant in all excluded material is that it will find a way to be included, eventually. Waste cannot ever be disposed of entirely effectively; its spectre never truly departs. It is hidden in plain sight, but as sound it endures, its presence always felt as a quantum of energy that has a kind of weight.[14] Even when processes are developed to, theoretically at least,

[14] It is common to describe sound and music in terms of 'weight', as either heavy or light. But recent findings suggest that it may actually possess mass: https://physics.aps.org/articles/v12/23 (accessed 08 July 2024).

permanently exclude 'unnecessary elements of sound', they never truly perish, as Maguire has demonstrated so emphatically.

Cox attempts to develop a material framework for the understanding of sound as a component of all art, in a manner that eschews convenient notions of representation and signification. In a way, uncompressing art. Here this explanation is expanded to include Whitehead's second level of symbolism, language, as a sonic phenomenon. The intention is to continue this expansive move and indeed to embed sound, not just in all aspects of art, but in all aspects of life, and most specifically within political economy.

In many respects, what Cox argues for is the need for a concept akin to a sonic image, where sound both serves as the ground for an immanent singularity, and as an animating force in terms of creative intervention. The explicit idea of a sonic image has been expounded by the artist, and 'private ear' Lawrence Abu Hamdan who proposes 'using the sonic imagination to trouble the image. Rather than sounds being forced to behave in the logic of the image [he saks], how can an image behave like a sound? What is politically at stake in making and seeing images that leak through and beyond their sensory and spatial frame?'[15]

In answer to this question, Hadman gives some pertinent examples of how sound can permeate the dominant visual realm to generate impact across a range of legal and political spaces, saying 'an image behaves like a sound when its truth value is derived out of its relational rather than individual qualities.' This links directly to Cox and his challenge to Kim Seth-Cohen and Brandon LaBelle, not to situate sound within an existing framework of visual art, but instead to infuse all art, and indeed all practice, with a sound (sonic) dimension.

Cox, like others discussed here, initiates a critique of postmodern thought and its rejection of essentialism. The rejection of essentialism, though correct in his opinion, comes at a cost he believes. It instates and supports a specific dualism that he wants to challenge. That being the distinction between an objective reality and its eternal deconstruction through language systems that are always unstable. For Cox, as it was for Serres, postmodernism with its tendency towards a deferral of meaning can never in the final reckoning do anything other than ponder and negotiate – like Roden's description of a universe without rules, there are only patterns. The challenge then is to extrapolate a system that can simultaneously accommodate both patterns *and* rules (as Badiou attempts in

15 https://infrasonica.org/en/audiblematter/thesonicimage.

'Democratic Materialism and the Materialist Dialectic'). Cox seeks to directly address this dichotomy recognizing that 'Contemporary cultural theory thus manifests a problematic Kantian epistemology and ontology, a dualistic program that divides the world into two domains, a phenomenal domain of symbolic discourse that marks the limits of the knowable, and a noumenal domain of nature and materiality that excludes knowledge and intelligible discourse' (2011 147).[16]

Cox neither wants to define music (as sound that takes numerous forms) as an 'abstract' material/natural noumenon, nor a purely cultural/rational phenomenon, but through a reading of Friedrich Nietzsche and Deleuze, for whom materialism is instantiated not as a duality but as a singularity where subject and object emerge from the same material source, as a mode of materialism that is both natural *and* cultural. According to such readings, sound, sound, art, and music should not be thought of as abstract simply because they are not conducive to formal representation. For too long, Cox argues, sound has been misunderstood in this respect.

> Music is thus conceived to be either sub-representational, a primitive eruption of desire and emotion (hence its suppression by moral conservatives from Plato to the Taliban), or super-representational, pure mathematics. Thus, Descartes (1961[1618]) could write of music that 'its aim is to please and to arouse various emotions in us' (p. 11) while Leibniz (1989[1714]) could claim that the beauty of music 'consists only in the harmonies of numbers and in a calculation that we are not aware of, but which the soul nevertheless carries out' (p. 212). (2010: 149)

Language of emotions (virtual/will) or language of mathematics; the sub or the super representation – the pattern or the rule, the chaos, or the accord? In reality music is all of these, as simultaneously a material coming to order and a straining towards chaos. This immanent, modal interoperability is however

[16] This should probably read '*some* contemporary cultural theory'. The dominance of postmodern ideas indicative of the linguistic turn is accepted – until Deleuze at least, as Cox recognizes. The approach developed here is not simply a re-enactment of postmodernism. This is because of the notion of a geometry of listening that will be explored later, but which for now is a mode of engagement that deploys a simultaneous compression and expansion as a process that does not seek permanent resolution.

Also, the 'Kantian epistemology' continues to resonate in contemporary theoretical debate, as exemplified by Roden's reading of Brassier. Kant proposed a synthesis of the positions adopted by Leibniz and Hume to produce what he referred to as transcendental deduction, and while this is not the primary concern of this book, it remains fundamental to the argument being developed and so will recur intermittently.

in contradistinction to the claim of Seth Kim-Cohen (2009) that to adopt a materialist stance in practice constitutes a return to essentialism. For Kim-Cohen any form of analysis that seeks to escape from the fluidity of cultural discourse, and the privileged ability of humans to execute it, constitutes a return to the dark ages insofar as hitherto inanimate 'unthinking' objects are afforded a degree of agency. For him sound must reside in the realm of rational human activity. And his is not the only counterpoint. Others pursue a dialectical reading of materialism that equally privileges humans as beings distinct from the objective world rather than the monistic position that Cox's rendition of Nietzsche demonstrates.[17] This form of materialist dialectics proposes a compositional process where sound and its description are engaged in a constant dialogue. Composers craft the unruly material of sound using notation as a tool – continually compressing and bringing to order while at the same time being challenged by the wild and expanded sonic terrain with its limitless potential. This 'creative' act is one that can provide a useful model for fields beyond music as will be discussed in due course. Specifically, the representation of sound through notation is important in the present context, and understanding the compositional process as a representational act of compression in an expanded universe is key to the political ambitions of this work. Whether via the deployment of dialectics or monism, the relationship between subject and object, the symbolic and the real, the positive and the negative, all need to be resolved. It is not however, as will be made clear, a question of either/or. Rather 'natural' humans and the 'natural' world coexist as an exchange and interaction of energy, information, and particles.

Hence musical composition, as an encounter between sound and its description, should be understood as a manifestation of materiality rather than abstraction. While careful to distance itself from the independent genius figure of the composer, the position adopted here also resists arguments that seek to remove (or are understood as seeking to remove) the act of composition altogether and replace it with an anonymous form of sound art that no longer requires human agency. Instead, it seeks to liberate and substantiate the real and the material by collapsing the subject/object dualisms while retaining a mode of creative potential free from 'the self'.

17 See Luc Dobereiner, in *How to Think Sound in Itself? Towards a Materialist Dialectic of Sound* (2014).

Luc Dobereiner, however, reads it in a different way. For him, like Kim-Cohen, the materialism of Cox (and by association the Deleuzian model on which much of his argument is based) is guilty of what he calls 're-enchantment'. Nature, he concludes, is granted the status of spiritual divinity in a manner that in turn serves to completely undermine the materialist intentions of the argument. For Dobereiner, it is the continual interaction, and the impossibility of resolution, between the composer and the sound material, that gives music and sound art, or music as sound art, its expressive energy, and its material presence. Drawing on the ideas of Slavoj Zizek and Jaques Lacan, to support his claims, he describes a gap between the real and the symbolic, the insurmountable tension between which is what constitutes materiality as a negative rather than a positive force.[18] It constitutes a negative dialectics where that is understood as a process of engagement that is perpetually under review, a project that is never complete, where the materiality of object and concept continually interact.

On one level this argument appears to be valid. The process of perpetual review is not in question. What is in question however is the perceived gap, or 'lack' between the real and the symbolic. What if concepts, as the bridge between the real and the symbolic as Brassier described them, are naturally occurring examples of compression that are generated as efficient, in varying degrees, modes of pattern formation and repetition? Not a deficiency in the concept and its relation (nonidentity) to the object but a positive process of constant renewal. Herein lies the problem with such a dialectical model with its constant splitting, discarding, and abstracting. Even if it is temporary and practical it still constitutes a problem insofar as you lose an important element of materiality in the gap that excludes that which is not immediately apparent and present in perception.

Dobereiner says: 'Every language or symbolic order contains gaps, exceptions, and impossibilities, for there is no total set of all signifiers, no complete symbolic order. Every attempt at a total symbolization remains incomplete. There is always a remainder, a residue, which reveals the inconsistency of the symbolic' (2014: 4).

This deficit is rationalized through compression, and it is precisely this gap that Maguire confronts in his *Ghost in the MP3*. The missing information, as a kind of sound shadow, is given a positive revitalized presence.

[18] Such dualisms are a kind of algorithm – dialectics as algorithm, a simple staged process that ends with a stable conclusion, that in turn forms the ground for a fresh challenge.

To further state the importance and inconsistency of this gap Dobereiner draws on the ideas of Karen Barad and particularly her notion of the cut, as a separation between the observer and the observed. His use of Karen Barad to support the role of the composer in the constitution of sound as real, however, does not necessarily tally with his use of Zizek's brand of dialectics. Zizek proposes a structural void, a gap that cannot be closed, and as such implies a kind of silence. It is a silence that is challenged by Cox's Deleuzian notion of a dense plenum where there can be no void, and where the cut (as described by Barad) is the result of a form of friction as noise. It is through noise then, as it emerges as sound and music, and not silence, that the waste and the residue, that Dobereiner himself refers to, are generated. Rather than gaps that cannot be breached, Barad's notion of the cut can be understood as the material formation of patterns among other patterns that intersect in a dynamic fashion where observer and observed are generated from quanta that are relative to each other and to the circumstances of their formation, a formation that divides and combines infinitely and at relative speeds to produce both consistent objects and inconsistent events. So rather than marking the divide between parallactic dimensions, as is the case with Zizek, the cut is dynamic and unpredictable, but always possible and positive. It is not a convenient cut in two, but a process of multiple cutting and healing.

At its most basic level then, the argument seems to revolve around the role of the artist.

> Sound and its perception come to be in processes that connect the emergence of the interiority of the listening subject and the independent exteriority of specific material sonic events and differences. The sound object presumes the existence of a stable form as an ideal gestalt recognised by the subject, and thus fails to acknowledge the emergent nature of transitory perceptual identities that are brought forth in the dialectic of the subject and the world. (Dobereiner, 2019: 219)

In saying this Dobereiner attempts to protect the status of the composer against Cox's perceived attack. It is argued here however that this is not actually what Cox is trying to do. He simply reconstitutes the ideal thinking subject within an immanent context that cannot be separated from material reality. To do so undermines static representational aesthetic practices and the political frameworks that they support. For Dobereiner, on the other hand, symbolic notation becomes a significant aspect *of* the real, rather than a simple representation. As such he delivers a rearticulation, via Zizek, of Hegelian logic and its attendant confirmation of the power of the intellect or the genius of the artist and with that come all the problems associated with dominant ideologies and canonical practices in the arts. While it does offer a way of including music and artists within sound art, it remains unclear as to why there is a need to re-enact dialectics to do so.

The argument ultimately invokes the question, is sound an object in its own right separate from a perceiving subject? There are many discussions pertaining to the ontological status of sound, and these will recur throughout this volume. Much of the discussion centres around the relationship between the listening subject and an ideal, or gestalt, sound object. Pierre Schaeffer described sound as an object independent of its source or origin that took material form through a correlative interaction with a listener. Others have argued for a more independent version of the sound object, or sound as thing, through a reconstitution of the Heideggerian distinction between object and thing.[19] This is an extremely interesting and relevant perspective, but one that ultimately falls back into Kantian deduction or phenomenology, and as such continues to prioritize the human subject with its sensorial gifts as the primary source of knowledge and as arbiter of the real.

And so, to another question, the representational or non-representational realm – which is real? The former is cultural, rational, and driven by language, the latter material, natural, and molecular (abstract because it is not knowable outside the sense making of rationality). Or should this be reversed? Through the interoperability of these two modes an abstraction occurs. But where is the abstraction? Abstraction is the non-representational, a real not open to experience through human rationality and language; but thought differently, representation as linguistic form is also abstract because it is not itself 'real' but rather a facsimile. In one sense the real is what humans can successfully capture and compress through the linguistic expression of sense data. On the other hand, the real is that which cannot be captured in this manner – a natural cosmological or quantum real that can only be speculated on. The significance of this will become more apparent in the next chapter where A. N. Whitehead's contribution is considered in more detail. Either way, the temptation to persist with dualisms and with dialectics can and should be resisted. For now suffice it to say, and without claiming to have the final word, it is important to state that

19 Dobereiner (2019), *Towards a Materialist Conception of Sound as Thing* https://www.cambridge.org/core/journals/organised-sound/article/abs/towards-a-materialist-conception-of-sound-as-thing/CC7D6F4E4BE25A01FDB6B08D101A5A6E.

the combination of ideas from across the range of philosophical arenas – but perhaps most pertinently from process philosophy and the philosophy of science, engaging their advocates and their critics – serves to underpin a commitment, as stated here, to a materialistic reading of noise, sound, and music where that materialism applies in equal measure to both subject and object.

Support for such a position can be underpinned by referencing another set of ideas. If our attention is turned away momentarily from the domain of continental philosophy, and towards a more analytical position, then further weight can be given to the argument being put forward. The analytical philosopher Wilfrid Sellars described the coming together of the manifest and the scientific image (as representation *and* will) in such a way as to accentuate the extent to which philosophers had come to deny the reality of perceptible things. This he believed led them to a dualistic theory prefaced on the divide between the subject and the object. By way of an alternative he describes the human body as a system of particles, and this includes thinking and feeling. These too comprise complex interactions between physical particles that constantly change state. This opens the possibility that the 'the self', the human as one being capable of carrying out multiple different acts, or improvisations as described by Brassier, can be replaced by a more systematic model based on scientific proofs.

This is remarkably similar to the argument that Nietzsche developed in relation to the spiritual modulation between nature and culture. What might be understood as a state of mind then becomes a state of matter, where the presence of an observing subject as material object can have a discernible impact on the laws of nature, as described by Barad (2007). Hence subject and object come into complex alignment and resonate as temporal events, which means that in terms of sound, meaning is entangled with its object and not separated as is the case with the visual register where distance is a feature of the relationship between signified and signifier. In sound the distance is breached, and the space is shared, combining through movement and resonance (as identified by Zuckerkandl). This movement, as described by Jean Luc Nancy, ensures the avoidance of static representation and closure. Instead of a phenomenological subject that is capable of ordering the universe into a set of predictable signs, Nancy posits a resonant subject, a self that is constantly in concert with the shifting nature of its surroundings through a multiple and complex set of exchanges.[20]

20 The subject of the listening or the subject who is listening (but also the one who is 'subject to listening' in the sense that one can be 'subject to' unease, an ailment, or a crisis) is not a phenomenological

The self, however one accounts for it, is nothing outside of the system of referrals and communications in an energy exchange, as both Brassier and Serres tell us. There is then, according to this position, no need to separate the symbolic and the real. The lack of closure 'in the economy of resonance and renvoi' that Nancy proposes prescribes a subject immersed in a chaotic universe, emerging, and unfolding, harmonic and dissonant, and always relative to the conditions of its existence. And importantly it means that the significance of composition, that Cox arguably puts into doubt, can be accommodated through the figure of resonance, and as will be explained shortly, through noise, itself a series of infinite referrals where the subjective composer and listener are themselves patterned as relative noise-events formed from vibrating energy.

Staying with the practice of composition, it is important to reference, by way of an introduction to concepts presented in what follows, the work of Pierre Boulez, who stated in relation to his own conceived method of composition: '*What we were doing was to annihilate the will of the composer in favour of a predetermining system.*'[21] For Boulez, tonal composition was a mode of compression that limited expression whereas his serial approach afforded an expanded range of creative options.

> According to the Gravitational Synaesthesia model, music can be explained as a series of interactions between particles (musical notes) and forces that link the particles with a range of intensity values. Thus, the evolution of music has involved, to a certain extent, the apparent lackening of mode-functional relationships, which, within limits, has resulted in atonality, involving all the tones for complete equality of their rights, and liquidation of whatever tonal center and gravitational pulls exist among them. (Galeyev, 1993: 77)

This describes both the tension between, and the simultaneous existence of, things that are understood as belonging together and things that are

subject. This means that he is not a philosophical subject, and, finally, he is perhaps no subject at all, except as the place of resonance, of its infinite tension and rebound, the amplitude of sonorous deployment and the slightness of its simultaneous redeployment – by which a voice is modulated in which the singular of a cry, a call, or a song vibrates by retreating from it (a 'voice': we have to understand what sounds from a human throat without being language, which emerges from an animal gullet or from any kind of instrument, even from the wind in the branches: the rustling toward which we strain to lend an ear). (2007: 21–2).

21 The idea of Smooth and Striated space – the musical model – as adapted from Boulez in Deleuze and Guattari's *A Thousand Plateaus*. Agency and the compositional act – counting patterns, rules, and the algorithm, serialism, somewhere between free improvisation and serialism, not counting and counting. There will be a more detailed discussion of this in the following chapter.

excluded. In his creative process, Boulez organized these interactions, following a series of compositional techniques derived from those created by Arnold Schoenberg in the early twentieth century. Schoenberg revolutionized the compositional methodology and the organizational features of music following the dodecaphonic technique. One of the more relevant aspects of this technique is that music has no centre of gravity anymore and therefore, there is no fundamental musical tone around which the rest of the musical material has to gravitate. According to Boulez, whereas 'classic tonal conception is based upon the view of the universe as controlled by gravitational attraction, the conception of serial music [derived from the dodecaphonic one] refers to the hypothesis of a permanently expanding universe' (Boulez, in Galeyev, 2003: 131).

The system of atonal serialism that Boulez developed sought to liberate music from its constraints, but somewhat paradoxically it did so by deploying a strict mathematical model. The system, as understood here, encapsulates a modulation between nature and culture where, even when annihilated, the will of the composer is still a system. As such it serves to underpin the argument that in noise, sound, and music can be found the code for understanding the simultaneous modes of compression and expansion that explain the emergence of all things as order in a chaotic universe.

In developing a mathematical system to enhance 'emotional freedom' Boulez also introduced to the world the terms *smooth* and *striated* to describe a free, unbound musical form and another based on counting, and these terms will become significant later when their adoption by Deleuze and Guattari is considered.

Variably, music must be understood in this context as a phenomenon that can be accounted for entirely via recourse to rational engagement, and where it is treated as a kind of aural geometry, or it can be consigned to a realm where only experience of it exposes those mysterious primal forces that cannot be accounted for but only inhabited. Either way, the Kantian notion of doubt (in our senses) remains. There is always the tendency to accept that there exists something other than what we can know either through direct experience or through the application of rationality and a priori models. Ultimately this inaccessibility renders such phenomena abstract and therefore real only in a theoretical sense. This has been a problem for philosophy since its inception. The figure of compression, specifically as a digital phenomenon in relation to sound, can offer a way of countering such lingering challenges. It provides a model for the systematic exchange and codification of data where contraction and expansion are relative aspects of a singularity that consists only of the real.

In this context what can noise, sound, and music teach about empirical and rational modes of being that wasn't adequately resolved by Kant? Kant forged a union between idealism and empiricism wherein sense data and its perception formed an unbreakable correlational chain. In the final assessment however, the rational subject in possession of innate reason was so significant a link in the chain in terms of bestowing the status of reality that the world *in itself* outside of appearance could never be authentically experienced. His synthesis operates in such a way as to preclude knowledge of the world *as it is*, meaning ultimately, we must settle for knowledge of the world *as it appears*. This is because of the unbreakable bond that conjoins the duality of rationalism and empiricism and accommodates both equally. The problem, where there is one, arises from this bifurcated starting position. In the synthesis both elements not only remain but are paradoxically strengthened, providing a kind of solution but one where the dual presence of the rational and the empirical means that they each in turn prevent the other from triumphing and hence the epistemological compromise. If instead we start from the singularity that is noise, no such compromise is required. Noise provides the multimodal ground for an immanent philosophy and insofar is it percolates through all the senses, removes correlational bind that limits the subject/object dualism.

This 'Kantian epistemology' requires some further consideration, and it will be considered in more detail in the next chapter. For now though, it is important to stress the necessity to think again our unwavering commitment to the nature/culture divide, or to that between empiricism and rationalism. For rationalism and empiricism, each designates knowledge as either theoretical or affective, an approximate a priori intellectual capacity or a primal response to material conditions – either way knowledge is limited. Even when Kant synthesizes the two approaches there is still a remainder, residue, an unknown. But 'natural' materiality is not confined to inanimate or non-human objects, and equally the non-essential and eternal undecidability of meaning is not confined to human rational culture. Culture is natural, and nature is cultural, and this becomes evident through the process of compression, the reducing down of bewildering complexity to manageable quanta of data. This process is both natural *and* cultural (and now technological). And as a process it includes essentialism, itself a mode of compression that seeks to establish boundaries around the characteristics and qualities of what a thing actually is. In so doing it is distinct but not separable from the postmodern text as a mode of expansion that allows such characteristics and qualities to be redrawn, rewritten, and reinterpreted as part of the ongoing contract between author and reader, where meaning and value are constantly renegotiated. As a process, however, it is unlike Wittgenstein's linguistic contract, that will be discussed later, where meaning and value are fixed to ensure reliable, effective, and meaningful communication (a kind of linguistic essentialism based on mathematical principles?), and where anything outside of its purview must be passed over in silence. It should also be contrasted with Heidegger's method which, while rejecting the essentialism of a thing, reinstates it in the final reckoning as truth (as a kind of essence) emerges from the critical weight of his questioning. Conversely, compression is a process of constant patterning where order forms before being compelled to de-from, driven by its own energy and motion. As a mobile dynamic process, it is better understood as a sound rather than an image, as noisy and illusive rather than static and regular.

The idea that sound is possessed of specific characteristics, of the kind described by Cox, which make it particularly adept at modelling and describing dynamic systems, is by no means new. Schopenhauer in his discussion of *representation* and the *will* assigned equally special characteristics to sound, and more specifically to music. It operated, for him, not as a form of representation but as an expression of what he termed *will*, a higher realm of truth not accessible through other modes of expression such as painting or writing. This is significant here as the move toward recognizing forms of expression that are non-representational is important and welcome. However, the division of experience into two distinct realms, *representation* and the *will*, is less helpful. Schopenhauer regarded music as having unique characteristics that echoed the mysterious virtual world, but his inherent Kantianism meant that he continued to demarcate this ultimately inaccessible metaphysical/abstract realm from that of the natural, physical world, the world of appearance and representation that was mediated and compressed through reason and the intellect. Real *and* abstract, the will fleetingly reveals itself through music, as a form of art that itself must ultimately remain 'unknown' in any rational sense. However, music exists not just as an expression of the *will* but also operates as a mode of representation, so it is entirely possible and indeed necessary that it be used as a model to understand both representation *and* will. Going further even, through noise, the distinction between *representation* and *will* can and should be eradicated altogether.

The world in all its complex incomprehensible enormity can be compressed into definitive intellectually consumable portions, and equally can impulsively spiral out of our control and comprehension. Both states coexist as compression and expansion simultaneously, present as order, chaos, sound, noise. This was recognized by Nietzsche who believed both the Dionysian, as the untamed (uncompressed) realm of the virtual will, where sound and music reside, and the Apollonian, as the formal realm of visual representation (compressed), were immanent to nature as well as to each other. Neither constituted an absolute or essential state. For him nature was a creative force, an artist in its own right, and where a human artist was present, it was a presence that was subordinate to the power of nature – as itself an expression *of* natural forces, either positive or negative. He described this situation in terms of 'artistic energies', which could emerge without the necessary mediation of the human being *as* artist. The key thing for Nietzsche, as described in *The Birth of Tragedy: Out of the Spirit of Music*, was that the relationship between the Dionysian and the Apollonian, music and the visual and textual arts, was not one of opposition but one of spiritual modulation – the former always animating the latter through a dynamic process of self-organization. 'The priority of sound and music in Nietzsche's philosophy, then, is not an aesthetic choice but an ontological commitment: the commitment to the primacy of becoming, time, and change' (Cox, 2011: 157).

This process of becoming does not operate in the same way as the postmodern text with its eternal deferral, but as an immanent energy field that, while never complete, does reveal itself to mathematics and science (and possibly to music), as numerous advocates of the speculative turn have elucidated.

If the 'spirit of music', as a Nietzschean concept, exists within all other forms of art and life, beyond representation and signification, then music itself, it must be recognized, is not exempt from signification. In terms of its representation through notation and more recently through the process of recording and digitization, music has been subject to a number of regimes of 'signification as compression'. Both musical notation and digital compression will be explored in greater detail in due course. The drive to bring sound and music into the significatory realm where, in the case of notation particularly, the eye continues to be privileged over the ear, was part of a wider set of circumstances that served to instantiate political and economic paradigms that were founded upon ocularcentric media practices and related technologies, as has been well documented by McLuhan and McLuhan (1988) and more recently by Friedrich Kittler (1999). Yet it would also be through technology that sound would come to be liberated (as will be discussed in the following chapter). Noise, sound, and music as temporal forms then both resist and attract signification and representation – neither polar state is preferable or more persistent, and it is because of this entanglement that sound serves as a reliable model for the analysis of all phenomena where the shifting ratio of compression and expansion does not resort to received dualisms but operates within a singular, material, and modal realm.

The field of sound art, like most academic fields, is populated by various positions that always seem to draw us back to the Kantian synthesis, phenomenology, and the role of the artist/composer and ultimately to the anthropocentric notion of agency. Christoph Cox and Ray Brassier have given us accounts of agency that are crucial in relation to the task at hand here. If Brassier did not give a full account of how pattern and rule intersect in his *Unfree Improvisation*, further investigation into his philosophical position reveals that he is acutely aware of this extant concept. In *Concepts and Objects* he says:

> We gain access to the structure of reality via a machinery of conception which extracts intelligible indices from a world that is not designed to be intelligible and is not originally infused with meaning. Meaning is a function of conception and conception involves representation – though this is *not* to say that conceptual representation can be construed in terms of word-world mappings. It falls to conceptual rationality to forge the explanatory bridge from thought to being. (*The Speculative Turn*, 2011: 47)

This rationality (as agency) is an ordering process or a compression of patterns into rules. Placing an emphasis on the ideas of 'access' and 'bridge' as they are deployed here by Brassier must be understood within the broader context of the thinking of A. N. Whitehead, and more specifically of Michel Serres's 'The Origin of Language' wherein the *we* is taken to be a plurality of subjective beings immanent to a complex thermodynamic system of information exchange and compression. These positions will be set out in detail as this volume unfolds.

For now, it seems appropriate to state that objects, since that is after all, all there is, do have agency, to a degree. The argument here is one of a flattened (but not entirely flat)/compressed ontology of democratic objects that recognizes the lingering adherence to the significance of human intervention and creative practice, free from the constraining concept of the 'self'. If at times speculative realism obfuscates the subject as agent of correlationism and defends the right of the world to exist outside of human perception of it, then through the figures of noise, sound, music, compression, and expansion, the subject (as object) is rehabilitated here. Placing the human (physical and metaphysical) firmly within a dynamic, sonic, material, immanent realm, restores moral and political responsibility without recourse to obsolete dualisms and dialectical logic. The subject is now an improvising being, an entity imbricated in systematic relations of information exchange where sound, as a model for pattern and rule formation, describes the system. It is drenched in noise wherein complex communication exchanges precipitate in such a way as to make indistinguishable the subject and the object but which still allows for account to be taken of agency, with chaos, if only rarely in the greater scheme of things, resolving to order and in the process constituting our lived reality – a small but no less significant accord that must nonetheless be situated within a greater field of potential and disruption. The confluence of matter/data/language/sound that forms our world should not be clung to as a sacred and permanent artefact however, for to do so foretells its own demise/death, but should be celebrated instead for its transitoriness. To acknowledge this is to enact a shift in perspective, a sensory perspective that can accommodate the ubiquity of imminent challenges without recourse to the tyranny of difference and the submission to the inevitability of nothingness.

2

Language as a mode of compression

Setting out from A. N. Whitehead's lectures on *symbolism*, *modes of thought*, and *the concept of nature*, this section moves through Michel Serres and his deployment of noise in a systematic analysis of language, to Wittgenstein – taking in music and notation, and moving into the digital present where compression is a recurrent feature. The key protagonists will be joined intermittently by various interlocutors, as the emergence of symbolic forms of representation is explored within a material immanent network of systems.[1]

Systems are often driven by what can be understood or described as a 'compression algorithm' with complex sets of circumstances reduced to common denominators, regularized and repeatable as a means of producing sensible, logical outcomes. As previously stated, this is not in itself a bad thing, but when such systems succeed in reducing the world and everything in it to static measurable (ontical) co-ordinates locatable in space and time they fail to capture the fullness of a universe driven by energetic exchanges and the confluence of events. So, a situation arises where to include risks our being overwhelmed and to exclude risks only partial access to the fullness of the universe. To counter this, better systems are needed, ones that are cognoscente of the process of their own formation. This is both a philosophical and a methodological undertaking. And importantly here, it is also an aesthetic one. It might immediately be asked in relation to this, are art and philosophy immune to systematic compression processes? I would suggest not. But they are processes of a different order in that they set out from a primary position of assemblage (akin to Boulez's serialism).

[1] The intention is not to make a specific contribution to wider scholarship as it relates to comparative studies of Whitehead and Serres, but to work through some of their concepts, mixing them in various ratios to find synergies, both resonant and dissonant. To this end, and in relation to Whitehead in particular, it has been decided to concentrate on those texts on *symbolism* (1985) and *modes of thought* (1968), that are related most directly to the themes under investigation here, rather than the more complete philosophical model developed in considerably greater detail in *Process and Reality* (1969).

Unlike merely functional modes of perception and simple recognition they seek a more complex mode of engagement, where the interdependency between simplicity and complexity is key to appreciating how overall systems and processes work. But what are the specific processes and systems that we are discussing? What is it we seek access to? For the most part it is systems of perception and knowledge creation where the aim is to dispense once and for all with the 'I', the subjective self, instead positing a reformed subject that submits to noise as an immanent constitutive material plenum. To do this means that beings, understood as events, are very different beings from the subjects inscribed by generations of thinkers. Equally the coalescence of such events constitutes very different social, political, economic, and aesthetic landscapes as they form across smooth and striated space. The processes in question consist of material and conceptual encounters between 'actual entities' or 'substances' (matter as energy) involving data/sense data transference, and processing/reconciling into concrete and symbolic forms. From a universe of almost limitless possibility emerge patterned formations as actual entities, including human beings and their sentient capacities (that are also events – matter and energy) – distinguishable but not separable from the ground from which they emerge.

The universe comprises elements that interact to form things. It is a wild and bewildering space. Yet within it patterns form. These patterns, groupings, and events encounter others in an 'infinite' choreography. Out of chaotic interacting matter life forms emerge. At their most advanced, the exchange of information as energy comes to organize itself into what we understand to be beings. As products of material interaction, they are embedded in a universe of things, yet their systematic mode of existence achieves such complexity that they perceive themselves as somehow distinct, transcendent. This sets up a dualism, object/subject that enacts a withdrawal of humans from their world, a withdrawal that forms a perspective or view. From this vantage point humans have measured, named, represented, everything in sight, fixing it forever as rational data that in various combinations is capable of not only explaining the universe but enacting it, directing it. But the system got out of control, began believing in its own spurious powers – a very familiar trope as we move into an age of artificial intelligence – but as will be discovered it is prone to hubris and to error. Indeed, hubris was born out of error, but paradoxically perhaps it is 'error' that will provide salvation. That which has been discarded will return as a corrective. It's time to get down from the vantage point.

Before starting in earnest, it is important to contextualize these systematic positions. The philosophical landscape has long fluctuated between rationalism and empiricism, idealism and materialism, and numerous other dualistic solutions, and the social, political, economic, and aesthetic implications of these have been significant. Central to this has been the Kantian notion of transcendental deduction. Kant attempted to align and account for the differences between Leibniz and Hume.

Leibniz advocated rationalism – the use of reason to account for reality without contamination of experience. Humans possessed, he believed, an innate ability to reason that, when appropriately formed into a system, could account for the world as it is (experience could be flawed but reason was pure). For Whitehead as we will discover, error belonged to symbolic reference, an aspect of a system of reason as abstraction or as argued for here, compression.

For Hume, who advocated for an alternative mode of empiricism, knowledge of the world was reduced to a point of view. The exercise of reason and the origin of ideas depended, for him, on experience, on knowledge acquired via the senses. So, all reality, for sentient beings, insofar as it is mediated is limited (an input error would produce an output error – knowledge would always therefore be compromised by perspective, always subjective).

Kant sought to synthesize these two positions: '*Kant wished to give an answer to the question of objective knowledge that was neither as absolute as Leibniz's or as subjective as Hume's* (Scruton, 1981: 21).

> Neither experience nor reason is alone able to provide knowledge. The first provides content without form, the second form without content. Only in their synthesis is knowledge possible; hence there is no knowledge that does not bear the marks of reason and of experience together. Such knowledge is however, genuine and objective. It transcends the point of view of the person who possesses it, and makes legitimate claims about an independent world. Nevertheless, it is impossible to know the world 'as it is in itself', independent of all perspective. (Scruton: 27)

How does it transcend a person's point of view? If it is through common acceptance of a certain logic appropriately communicated through language, then claims to objectivity are perhaps made in haste. General acceptance of knowledge is not the same as it being objective. Rather it is, referring back to the *Ghost in the MP3*, a common denominator logic as epistemological compression. Whether possessing innate rationalism or empirical data sensors (or both) human beings retain a privileged position in terms of enabling as well as limiting knowledge. As Steven Shaviro says of the conditions of possibility created by Kant's thought,

> Nearly all contemporary philosophy is premised, as Lee Beaver shows in detail, on a fundamental antirealism; it assumes one version or another of the Kantian claim that 'phenomena depend upon the mind to exist. Such philosophy denies the meaningfulness, or even the possibility of any discussions of 'things in themselves'. Modern thought remains in thrall to what Harman calls the idea of 'human access', or to what Meillassoux calls correlationism. It gives a privileged position to human subjectivity or to human understanding, as if the world's very existence somehow depended on our ability to know it and represent it. (*The Speculative Turn*, 2011: 280)

So how can we move beyond this Kantian hurdle? Can knowledge be separated from the subjective condition of the knower and what happens if the subjective condition of that knower is constituted from individual objective factors in combination/relation – as events? This is what Whitehead proposes. The task here is to elaborate on how this proposition can be established to support meaningful systems of contemporary engagement. Human beings, their ideas, their art, their science, their architecture, their technology, and their social forms of organization are all systems wherein noise and chaos resolve to some kind of order. These chaotic systems are dynamic, their spatio-temporal dimensions relative, with some assuming a degree of 'permanence' while others demonstrate transitoriness. In reality all are in motion, each relative to the other. A critical understanding of this dynamism affords new understandings of our relationship to (or within) the material world and to history. As argued for in the previous chapter, noise, sound, and music, as part of a more general challenge to ocularcentrism, become important aspects in terms of supporting contemporary knowledge frameworks as the need to move beyond human limitations intensifies. Such frameworks should serve to support systems of engagement that are dynamic and responsive, ones that accommodate diversity without recourse to a tyranny of difference (more on this later) whereby the inclusion of everything sustains pandemonium and impedes our understanding of particulars.

A key system in the evolution of advanced organisms and their associated languages is symbolism. History plots both the superficiality and the increasing complexity of symbolism as a system of perception and representation. Transitory, human gestures beginning with grand 'emotional' responses evolve over time to constitute more considered, complex, and analytical modes of enquiry. As symbols, architecture, heraldry, and ceremonial events have given way to deeper, more indispensable symbolism of language, and yet deeper still in the form of pictorial representation (art). It is not as simple though, as a pure biological cause and effect response – a reaction to an external sensory (empirical) input. Symbolism evolves into extremely complex communicative systems that operate at different levels with varying degrees of sophistication – not just patterns but rules, to refer back to Brassier. Whereas science can account for many cause and effect relationships, the complexity of the subject/object relationship escapes its explanatory powers – for now at least. The concept of the human subject itself is imbricated in this process, emerging from the annals of continental philosophy to finally establish itself as existing at a distance from the rest of the material or objective universe. Alfred North Whitehead is invaluable in terms of our attempt to breach this distance, which is a key aim of this book.

The evolution of symbolism has continued since Whitehead's time of writing and now includes the digital mode, itself an evolution of his description of the mathematical mode as an important facet of symbolism. Yet rather than affording answers it further complicates the question of whether symbols are material objects (real), or whether they are more akin to abstract processes of ideation (and therefore virtual). If they are processes of ideation forming abstract representations, does this require a phenomenological model of perception wherein the subject is ultimately privileged in terms of information and data and its processing into the form of a symbol? For Whitehead the perceiving subject (or superject) as advanced organism is significant, but not necessarily privileged in a world of complex interactions (prehensions). Rather than existing at a distance, Whitehead places humans as immanent to their environment, emerging like all other events and actualities from a process of interaction. Likewise, the real and the virtual status of entities in discourses around digital ontology should be understood as immanent rather than transcendent.

Whitehead presents the idea/theory of a system or process that continually moves and evolves, and which can exist in both simple and complex forms. It would seem therefore to be well suited to addressing the digital present.[2] It equally accommodates sonority, noise, sound, and music. Even when

2 See S. Shaviro, *Without Criteria* (2009).

Whitehead discusses art (as predominantly pictorial) he does so in such a way as to foreground its functioning not simply as an act of immediate recognition and fixed representation, but of hesitation, contemplation, and dynamic emergence. This predates and adds weight to the arguments in the previous chapter where sound is understood to pervade the entire aesthetic realm. Though he rarely engages specifically with sound and music, his ideas retain a sonic dimension as will be elucidated in more detail further in this section.[3] The process he describes can best be understood/inhabited/experienced through sound. Insofar as it details inevitable change, it also has an inherent social and political dimension, and how this is dealt with is one of the key questions of our time.

For a more effective assessment it is useful to examine Whitehead's text in more detail.

Symbolism is Whitehead's response to Hume and Kant and is positioned here in relation to the previous discussions around the subject/object dualism and transcendental deduction. The impact of redefining the subject/object dualism in relation to symbols and how they work is significant. What follows is an attempt to inhabit the layers of integrated systems and processes within and between objects as subjects and subjects as objects operating on the basis of a relatively (but certainly never entirely) 'stable division' that distinguishes (in terms of modes of existence) without separating. The idea of relative stability develops throughout this chapter and can be enlivened through the invocation of noise, sound, and music. Things that are emerging and in process rather than fixed both exist and are in a state of transition, where sound as both object and emergent phenomenon becomes variously music that is recognizable and recurring, stable to a point, but always relative to its environmental and technical circumstances. It is smooth and striated, never absolute but entangled in immanent relations to itself and to the world.

Whitehead and symbolism

Medieval symbolism was a grand expression of wonderment and awe, perhaps even an act of submission in the face of a bewildering universe, like God, heaven, and other human responses to the mystery of the 'beyond'. From the reformation

3 Whitehead does address the notion of audition briefly in *Process and Reality* (1969: 272–4).

onwards attention turned to detail and the apprehension of ultimate facts. This change constitutes a process of breaking things down into their constituent parts, according to a scientific paradigm that seeks to get the world in order – to make it make sense, according to the senses. It largely finds its expression in language, as a mode of existence and enquiry that takes many forms. For Whitehead then, symbolism operates at a number of levels. It can consist of large-scale expressions of human achievement that are arbitrary, superficial, and subject to historical changes in fashion, and then there are those more significant symbolic forms like language that are almost essential to our existence. Language itself he says can be further divided into that which can be written or spoken, and the pure written-only language of mathematics. There then comes an even more significant and fundamental kind of symbolism: lived perception and experience of our world. In each of these modes, compression is apparent as complex potentialities are condensed into distinct concepts and representations. The intertwining of the mathematical mode of language and that of lived perception is currently evident in the digital landscape, and its apparent bifurcation into the real and the virtual. Whitehead, among others, helps us to reconnect these bifurcated realms and to conceive of a singularity wherein complexity and distinction do not constitute options or choices but are integral and relative aspects of a process of becoming. Details of how this works in practice are discussed in later sections.

The relationship between the substantial physical world and a more focused 'attention to detail' as it is rendered into symbolic form has itself been present in many arguments and described in many ways – as physics/metaphysics, noumena/phenomena, Being/being, Dionysian/Apollonian, will/representation, pattern/rule, chaos/order, noise/music, smooth/striated, and so on. And it endures. What remains prescient, but which has not to date been adequately settled, and most probably won't be here, is the way(s) in which these different modes connect/combine/co-mingle/resonate. Making a contribution to this debate however will go some way it is hoped, to allowing us to comprehend the intermodal process that constitutes materialism, the object, and speculation (as abstracted linguistic mode). To do so solely, or predominantly through a visual lens tends towards fixity, permanence, and universally applicable laws (scientific and social) whereas greater acknowledgement of the significance of sound allows for fluidity, transience, and relativity.

Symbolism, as it moves from the universal to the granular, is a mode of compression that brings the world into some kind of order. I say 'some kind' because there are many levels of order, and all are contingent on any number of factors. Language as it orders through spoken, written, and algebraic modes, is a means of organizing sense data into comprehensive meaning, and in doing so it creates/brings into being, a kind of waste or residue (all specific characteristics of a thing become subservient to the name given to the thing), as that which doesn't fit, or that which cannot be named or *put into* words. It categorizes and normalizes, as Foucault tells us so brilliantly at the beginning of *The Order of Things*.[4] In the main it works and is incredibly efficient. Most of us conform to it most of the time (it forms a relatively stable division). Yet there are exceptions (and these exceptions will come to be very significant in the argument that follows). Certain people, artists for instance, can be trained not to compress data but expand it, and this means we need to consider whether art is a specific type or mode of sensory engagement (prehension) that can reveal to us the very process that Whitehead proposes, and which compression is designed to obscure. If so, then sound art/practice (as a significant factor in all art, as established in the preceding chapter) may operate at yet another, deeper level. As noise, sound, and music it can compress and expand in equal measure. By referring back to Whitehead and the idea of process as an evolutionary and transitory energy field, knowledge, its acquisition and communication can be understood as having dynamic sonorous qualities that are distinct from the perceived permanence of the visual field.

The different stages of language that can be historically traced constitute varying stages of compression – each a reduction in scale, each less ornate, but potentially more effective and efficient. Written and spoken languages, insofar as they still require direct participation by an actor (mostly human), retain the potential for nuanced interpretation and adornment, and with that misunderstanding. The algebraical symbols however expunge this weakness

[4] This book first arose out of a passage in [Jorge Luis] Borges, out of the laughter that shattered, as I read the passage, all the familiar landmarks of my thought – *our* thought that bears the stamp of our age and our geography – breaking up all the ordered surfaces and all the planes with which we are accustomed to tame the wild profusion of existing things, and continuing long afterwards to disturb and threaten with collapse our age-old distinction between the Same and the Other. This passage quotes a 'certain Chinese encyclopaedia' in which it is written that 'animals are divided into: (a) belonging to the Emperor, (b) embalmed, (c) tame, (d) suckling pigs, (e) sirens, (f) fabulous, (g) stray dogs, (h) included in the present classification, (i) frenzied, (j) innumerable, (k) drawn with a very fine camelhair brush, (l) *et cetera*, (m) having just broken the water pitcher, (n) that from a long way off look like flies'. In the wonderment of this taxonomy, the thing we apprehend in one great leap, the thing that, by means of the fable, is demonstrated as the exotic charm of another system of thought, is the limitation of our own, the stark impossibility of thinking *that*. Foucault, 2001: xv.

'doing your reasoning for you'. Wittgenstein would strive to assign to 'ordinary language' (written/spoken) the same degree of efficiency.

Whitehead says:

> There are deeper types of symbolism, in a sense artificial, and yet such that we could not get on without them. Language, written or spoken is such a symbolism. The mere sound of a word, or its shape on paper, is indifferent. The word is a symbol, and its meaning is constituted by the ideas, images, and emotions, which it raises in the mind of the hearer.
>
> There is also another sort of language, purely a written language, which is constituted by the mathematical symbols of the science of algebra. In some ways, these symbols are different to those of ordinary language, because the manipulation of the algebraical symbols does your reasoning for you, provided that you keep to the algebraic rules. This is not the case with ordinary language. You can never forget the meaning of language, and trust to mere syntax to help you out. In any case, language and algebra seem to exemplify more fundamental types of symbolism than do the Cathedrals of Medieval Europe. (1985: 2)

Such 'more fundamental' types of symbolism are increasingly significant as the entirety of our existence is subjected to code as algebraic representation. The extent to which it does our reasoning for us is an issue of great contemporary relevance and will be considered throughout what follows. It certainly does it with us as Ryan Patrick Maguire's *Ghost in the MP3* has so elegantly demonstrated. Just as we naturally filter out and reduce the capacity in terms of the sounds we perceive, so the algorithm can isolate what is deemed to be unnecessary. As such Maguire's work gives us unprecedented access to the system of symbolic representation that Whitehead describes.

Symbolic representation in all its modes must be understood to have limits. In fact, representation is by its nature a limiting device. The striving towards greater capacity, accuracy, or reach in terms of representation is flawed so long as it attempts to create the perfect map (Gaiman, 2006). In certain immersive, spatial, and audio environments for example, representation strives to become so *like* the real world that we may as well resolve to make do with the world itself. But digital representation (as mathematical linguistic mode) doesn't quite work like that. Rather than producing an actual model or twin, it forms a compressed facsimile that reduces reality to a comprehensible and accessible state that precludes all of the difficulties of scale that prevail within the real. So, representation in this mode is compelled to omit or reduce, and while this produces clear benefits and efficiencies it also poses challenges. What follows is an attempt to address some of those challenges.

The digital mode of representation, though a significant advance in algebraic (and algorithmic) reasoning, retains as a central feature an important characteristic of all other modes: the ambition to expand and the compulsion to compress – now with much greater speed and accuracy. To better represent though is not the same as more accurately representing. Qualitative and quantitative factors and the interaction between the two are extremely important in any process, as brilliantly described by Whitehead, and they shift in relation to each other constantly. This process, or processing, is at the heart of the debate around speculative realism and the drive to situate knowledge of the world outside of human perception of it: to establish an objective real that does not require subjective verification. Yet the role of human beings within this process remains central to the task at hand here, because to adequately account for a world independent of human perception of it does not take them out of the equation altogether. The diminution of the 'self' does not equal the elimination of the subject. Even if that was possible, to do so would only serve to absolve them for their failings. A more appropriate course of action is to, as Brassier does and Whitehead has done, position them as immanent to both the problem and its solution.

When the world is 'put (by humans) into words'/pictures/sounds, code, or even material objects and configurations (buildings/cities/nations) as a form or mode of symbolism, a degree of compression is involved. For Whitehead compression is manifest in symbolism as a mode of representation or a stage in a process that is distinguishable from but not unrelated to the pursuit of 'ultimate facts' (qualitative and quantitative), and it comes in numerous forms or modes. To summarize,

- Grand, sweeping symbolism as an expression of wonderment and awe.
- Language as a symbol, words written or spoken that generate meaning.
- Mathematical language as symbolic – with an internal logic and syntax (to this list is added: the digital, as a subset of the mathematical).
- Perception and the staging of the symbolic – this being the most significant mode.

Perception happens at a level that is often beyond conscious recognition of the process involved (except perhaps in the case of a trained artist). Adherenc and reliability draw on past experiences of perception as a kind of repetition (perhaps not the Deleuzian kind). It is simple/natural rather than a 'high grade' act. Simple but not insignificant. 'It is not mere tropism' (no 'mere' tropism, but a tropism nonetheless). This 'simple' mode of engagement, that Whitehead calls *causal efficacy* is fundamental insofar as it applies to the interaction between things, and is the basis, in that it applies to some degree to interactions between *all* things, for subsequent symbolic acts in the mode of presentational immediacy. Steven Shaviro reminds of us of this saying:

> Western philosophy since Descartes gives far too large a place to 'presentational immediacy', or the clear and distinctive representation of sensations in the mind of a conscious, perceiving subject. In fact such perception is far less common than what Whitehead calls 'perception in the mode of causal efficacy', or the vague (nonrepresentational) way that entities affect and are affected by one another through a process of vector transmission.[5]

To engage with the world through language as representation then is to start at the wrong end of the process. Once symbols are established the process is no longer accessible to experience and an important mode is lost.

Causal efficacy and *presentational immediacy* will be discussed in more detail presently.

In the immediate term however it is important to note that presentational immediacy does not always merit the status given to it by Kant – a status that underpins the correlation between object and subject in the formation of the real. Such a position has led to an overstating of the role played by a rational human subject in the production of knowledge. Instead, greater attention needs to be paid to the countless universal and actual interactions that occur and that are not compressed into a human version (linguistic) of reality. Despite this, the passage from the perception of say, coloured shapes, to the recognition of a physical body, and the formulation of a response to that recognition remains a significant 'characteristic of advanced organisms', and such a process as this attains even greater significance when understood in terms of its being enacted by high-grade organisms such as humans. Human perception is important therefore in ways that primitive perception enacted by lower-grade organisms is not (Harman may disagree). Whitehead wanted to know how this process had

5 http://www.shaviro.com/Blog/?p=1274.

developed in humans as the highest grade of organism, but in a way that did not assume the subjective privileging of reason.

Simple recognition of a represented object, though, is in itself not a particularly 'high-grade act'. Whereas artists are trained to pause and consider colour, shape, and form, most people move straight to immediate recognition as an almost seamless act. It is the artist who is trained *not* to leap straight into perception. No training is required to refrain from the elaborate thought process of contemplation. Hence, this process of recognition is a basic mode of compression, a putting in order for the sake of calm, in the face of a chaotic torrent of data that is presented for consideration. Only later in the process, and only on some occasions, is this data reflected upon in any great detail. So it is that we turn to the artist, philosopher, and scientist, in their varying modes, to produce 'highest-grade' responses that are not solely dependent on the immediacy of recognition but expand into the realm of possibility.

The idea that reason and logic serve to bring a coherent and natural order to chaos is a well-established one, and while this is often the case, and while it is a brilliant and efficient system of compression, what reason and logic also do is to impose certain very specific patterns on the world through a variety of processes of selection and rejection, mediation, translation and design, that are too often definitive and fixed. What Whitehead, Wittgenstein, Zygmunt Bauman, and Michel Serres, among others do, is to demonstrate for us how compression and expansion, as instances of *the repetition of difference through resistance* (things in themselves enduring in space-time without surrendering to the static command of representation) embodied by artists, philosophers, and scientists/mathematicians combine to formulate a reconstituted mode of engagement with the world that is less reliant on the sovereign primacy of the rational subject.

In pursuing this course, a distinction must be made between symbolism and direct experience and between the fallible and the infallible. Symbolism is too often presented as a generally reliable feature of advanced organisms – that is, reliable in terms of the relatively stable relationship between the symbol and the world that is represented by it. In reality the potential for error remains. The process may be simple/natural, but it does not always engender predictable cause and effect (*causal efficacy*). Advanced organisms do not always respond in the way that might be expected of them. The (contested) task of reason it can be said then is to distinguish between truth and error. This process, the processing of reason, occurs at the intersection of direct recognition and symbolic reference. The latter must always satisfy the criteria of the former and ultimately refer back to it, and in predominantly visual modes of representation this is relatively stable, whereas in its audible mode things are less certain. To grasp the fullness of this process a more detailed assessment is necessary.

Presentational immediacy is a mode of experiencing that puts simultaneous things in order, arranging a torrential data stream into momentarily manageable segments. It is the direct recognition of objects, conscious or not, the 'projecting' of a picture derived from sense data (Hume calls this 'impression') and proceeds from an immediate recognition of that data. What one sees is what things are at that moment (a singular interaction as opposed to Kant's dualistic description). The objects perceived 'at that moment' and arranged into some kind of spatial order (they exist relative to our perception of them) with qualitative characteristics (colour for instance) to form a momentary account of the world that is present *through* perception and that is in the world itself, *actual*. There is no more fundamental objective real than this experience. As Isabelle Stengers has said, 'neither simultaneity nor coloured things exist outside of visual perception, but they translate a cerebral trust in a community of enduring beings that actualize a separation between space and time according to independent modes' (*The Speculative Turn*, 2011: 400).[6] It is a reassuring process that abstracts from the incessant and dynamic process of all things changing all the time, to ensure a degree of stability, familiarity, and logical expectancy, as cerebral trust. This trust is the basis of the eventual synthesis between presentational immediacy and symbolic reference as a mode of compression. It is though fallible.

What you see is what there is. It is in the interpretation of the data that potential problems arise (as in an optical illusion, where all the information is visually perceived, but the interpretation is unstable). More generally it is the emphasis on the visual and seeing that is problematic. The compulsion to bring to order a chaotic simultaneity occurs at the expense of aural sensibility that is better able to account for the dynamic and transitory nature of reality.

The percipient participates in the process of recognition insofar as perception *of* something rather than the something itself is (re)produced as a relation. This is what distinguishes it from causal efficacy. Colour and perspective, for example, are processed as distinguishable, abstracted but not separable, actualities, as percepts become concepts. Colour and spatial relationships/perspectives are not

6 It is not that objects, or the simultaneous and contemporary arrangement of objects, don't exist if we don't experience (see) them, but that they do not conceal their true selves from our visual perception. They can be seen. They show themselves as they are at that moment.

disembodied concepts that are projected onto the world but are experienced in themselves at that moment. One does not store the code for a colour and then apply it, but rather experiences *what is* through a relational process involving a degree of qualitative recognition beyond the most simple awareness that something is present.

Appearance (as presentational immediacy) is then a relative process of information exchange with equal weight attributed to the qualities of the object and our sensations that form through the processing of sense data. This relationship can only be disrupted (isolated and distinguished) by abstraction. Until then object and percipient are locked together in a spatially extended process wherein the percipient is also object. This simple mode of perception is meaningless, a pure recognition where qualifying concepts have not yet (if indeed there is temporal latency at all between causal efficacy and presentational immediacy) connected to experience. It is a physical exchange between things in the world. Presentational immediacy then combines with symbolic reference to produce integrated and contemporary temporal modes of experiencing. Once these phases are completed and processed into concepts an abstraction occurs and a mode of representational engagement with the world begins. Hence abstraction is a mode of compression – but still a very real one. The actual entity, or the *what* (object) that is perceived as abstraction (percept), and their relationship constitutes a natural phased process. The idea, supported here, is that abstraction is natural, and proceeds as a mode of compression: 'Abstraction expresses nature's mode of interaction and is not merely mental. When it abstracts, thought is merely conforming to nature – or rather it is exhibiting itself as an element in nature' (Whitehead, 1985: 26). The object and its abstraction as percept are forever entangled, moving as they do through a process that tends towards conceptual analysis. Sense data participates as an element of synthesis in a process that is progressive and incremental insofar as the percipient does not experience sensation as if for the first time every time as presentational immediacy, but rather comes to recognize it as familiar (but not as a code to be projected). So, synthesis where presentational immediacy fuses with symbolic reference enacts a degree of compression or efficiency within the process as it becomes ever more systematic. The evolution of the process into a functioning system leads to formal language as a mode of representation in a practical sense – the representational code (word or memory) for a colour is not the same as the experience of the colour itself (though as we have seen they are forever entangled), but it stands in for it in absentia (both the original object and the symbolic reference exist as actual entities) and means that we don't have to return to the original source each time for clarification, though if we do the object is again prehended in its actuality (as a process now aided and enhanced by abstraction/compression).

Even if every organism perceives the actual differently (by however small a degree), advanced organisms through this process of presentational immediacy, moving through symbolic reference, are bequeathed by nature, the ability to compress the vastness of reality and the universe into modes of language that can account for and communicate that reality without recourse to subjective phenomenological mediation. Humans as events take their place in the universe without the necessity for them to act as a conduit for its being. Humans are not the arbiters of reality but are themselves subject to it in the same way as all actual entities. As such mental faculties derive from nature and are themselves natural.

Ultimately, perception in the mode of presentational immediacy has its limits in terms of it forming the basis of symbolism and ultimately linguistic representation. Whitehead says that metaphysics requires a 'reformed subject' – one that is not separable from the objective world. In this sense language as a mode of symbolic representation (as compression) is of only limited use – 'language, in its ordinary usages, penetrates but a short distance into the principles of metaphysics. Finally, the reformed subjectivist principle must be repeated: that apart from the experiences of subjects there is nothing, nothing, nothing, bare nothingness' (Stengers, 2014 359).

This is what makes Whitehead's contribution so significant and important here. It allows for, or rather insists upon, the recognition of the subject, as actual entity, within an immanent system of relative states of being where all elements interact in a universe that is always changing. That there might exist something that cannot be experienced is what constitutes a nothingness. Reality resides within and not outside experience. The fullness of the universe in all its expanse communes with compression as a simple ordering mechanism in modal ratio to constitute the real. The now 'reformed subject' is not a withdrawn entity, but one that embodies reality through a number of modes of experience. If one of those modes delivers the 'ordinary usages' of language then it is imperative that different modes and usages be identified, ones that produce 'extraordinary usages'. They are sorely needed. Creative practice, art, sound art, philosophy, and science in various combinations must deliver these new forms of knowledge, and renewed articulations of the real in a world where error and limitation (often through omission/compression) within a field of symbolic reference, that supports current social, political, and economic configurations, too often prevail.

All of this means that the abstraction of source material into code/language allows for comprehension, consideration, description, and communication without recourse to direct experience. As such it is understood here as a mode of compression rather than innate rationality. Though prone to error it also gives rise to creativity as a form of expression. It does not constitute an *instead of* (as resemblance) but a process of ordering (compressing), with the real finding a way of describing itself through pattern and rule to attain a relative stability.

Referring to Aesop's fable wherein a dog drops a piece of meat while snapping at its reflection in the water, Whitehead says: 'We must not however, judge too severely of error. In the initial stages of mental progress, error in symbolic reference is the discipline which promotes imaginative freedom. Aesop's dog lost his meat, but he gained a step on the road towards a free imagination. Thus, symbolic reference must be explained antecedently to conceptual analysis, although there is a strong interplay between the two whereby they promote each other' (1985: 19).

Whitehead chooses to allocate to the realm (mode) of 'mental activity' experiences that combine both percepts and concepts. Ultimately though mental activity is also physical. Unlike Aesop's dog, however, some entities are 'purely physical' and cannot make mistakes. They are entities that interact within the mode of causal efficacy. This is an enormously significant point. The classification of error and mistake reside in the mode of symbolic reference and its synthesis with presentational immediacy. 'In short, truth and error dwell in the world by reason of synthesis: every actual thing is synthetic: and symbolic reference is one primitive form of synthetic activity whereby what is actual arises from its given phases' (1985: 21). Purely physical actualities engage in a different kind of synthesis – without error. Yet humans are quick to designate error and mistake here too: mostly in the form of normative values – sometimes for reasons of efficiency, but at other times for personal gain and domination, as an expression of political economy and aesthetic value. This is, of course, in itself a very good example of Whitehead's position.

As distinguishable from purely physical synthesis, symbolic reference requires a more active engagement on the part of a percipient, as has already been stated. 'We must conceive perception in the light of a primary phase in the self-production of an occasion of actual existence' (1985: 8). The active participation of a being in the conjoining of the symbol and its meaning, it must be stressed, confirms a materiality of perception that closes the gap between the world as it is and the world as it appears to be. For Whitehead the coming into appearance is actual. Yet as a process it remains fallible, and this fallibility must be accountable in terms of moral responsibility. Human beings play an important (though not essential) part in a process in which they are inculcated. Humans become factors (actors, or actants), albeit significant ones, in a process that reduces them to a stage in a process that both requires and undermines (in terms of Being) them at the same time. Here the physical and the metaphysical entwine in a relational exchange as percept and concept. 'Thus in fact there is no proper line to be drawn between the physical and mental constitution of experience' (1985: 20). They are in communion, both dissonant and resonant, via a process of energy exchange.

All actualities interact to form patterns. Whitehead describes this belief as an incessant process of interaction where everything is connected to everything else. But Whitehead's process is not a case of transcendental deduction. It does not distinguish existentially between noumena and phenomena. The objective entity and its perception both constitute actual entities, and as such they correspond, prehending each other with varying degrees of perceptive sophistication, in a constant interchange of energy and information that coheres momentarily as meaning (signal) emerging from noise. Energy emitted from an object provokes an energetic response from a percipient as the different frequencies interact. In this respect Whitehead is a significant antecedent of the speculative realist movement. As such it is worth dwelling here momentarily to consider why the differences that exist between Whitehead and Graham Harman, as an advocate of speculative realism, are significant.

An assessment of the differences between Whitehead and Harman reveals the political significance of a model of thinking that is not reliant on dialectical modes of thought and the separation of the real and the symbolic. The interconnection of all things in Whitehead's work opens up the idea of an expanded field of consideration wherein all aspects are significant. For Harman the ultimately closed impenetrability of things, despite acknowledging their reality, means that speculation is as close as we can get to knowing them. Both give rise to different political modes. For Whitehead everything is connected through a process of emergence. It is a vision of a universe in constant flux where encounters, actions, and reactions somehow resolve to moments of order as events form into discernible patterns that mix with other patterns, sometimes coherently, sometimes not. Everything that exists as an actual entity can potentially combine with and/or encounter everything else, and it is these encounters and combinations that constitute a reality that is never dependent on a (as a priori) subject for its validation. Such an objective reality as described by Whitehead is welcomed by Harman, but for him, rather than everything being connected as part of a monistic process of becoming, independent objective entities are in the final reckoning removed from one another. Despite their interaction their true self is never revealed. They are effectively hiding in plain sight – present but anonymous. For Harman the universe is constituted of real actual entities that are distributed in such a way as to conceal any ultimate meaning. Hence it is a reality that can only be speculated upon. The importance here is that while Harman's model conforms to the ocularcentric tropes of visual perception and hiding from view, Whitehead's alludes to a universe of vibrating energy and frequencies that might better be understood from a sonic perspective.

Conscious that we have diverted a little from our path but with the dynamism of Whitehead's process fresh in our minds, let us return more directly to the task at hand.

Whitehead and music

The contribution that Whitehead makes in terms of his process philosophy, while not specifically related to sound or sonic perspectives, does invoke, through its inclusion of movement and temporality, a sonic sensibility. He is not concerned with instants in time or specific points in space, but rather the emergence of 'events' in process, relative and temporal. His position, vacillating as it does between monism and pluralism, describes a universe in which events as actual entities emerge from an energetic concrescence into ordinary and eternal objects – things and symbols, things as symbols and symbols as things, as elements of orchestration on a universal scale.

Referencing F. David Martin's *Art and the religious experience: the language of the sacred*, Richard Elfyn Jones brings Whitehead directly into the musical domain. He quotes Martin to emphasize the belief that music is perceived in the mode of causal efficacy wherein the presentational immediacy of hearing present tones always evades our capture as the promise of what might come next prevails and exemplifies the very nature of process. Music is the sound of the process at work (and it comes in many forms). If presentational immediacy serves to fix and order it is largely impotent when met by the fleeting transitoriness of sound.

But this is certainly not a straightforward process. To understand the perception of sound in a Whiteheadian manner we are urged to think about what happens for instance when the sound of a door shutting is perceived. On one level the door is prehended by a listening subject within whom mental processing takes place to establish that the sound came from the door and noting *that the effect of the sound is caused by door shutting.* This is perception in the mode of causal efficacy. If, however, we want to understand this process from a transcendentalist perspective, its presentational immediacy as a symbol, it is imperative that the sound of the door shutting is abstracted from its source, the door. This describes the acousmatic process where the listening subject is separated from the objective sound source, so that the process of listening can be understood in isolation, as an abstract phenomenon. This is a Kantian position and has been dominant in continental philosophy for a great deal of time as has already been established. Whitehead, and others who have followed, challenge this position. The ontic (the actual entity) and the ontological (the process of its becoming), though distinguishable, are never separable. The same applies to noise, sound, and music.

The symbolic nature of sound in an acousmatic environment is a technological configuration of extra-dimensional prehensions that occur as naturally as they do artificially, and are part of a sensorial process that evolves to constitute high-grade beings wherein data processing, storage, and communication involve compression (symbolism) to avoid the need to experience all encounters (prehensions) as if for the first time – a kind of memorial, linguistic (symbolic), efficient (mostly) process. Yet, like in the Whiteheadian model overall, it is important not to overstate the processing of sound sensations and the role played by the subjective percipient. To do so risks overestimating the significance of the role of a subjective rational being in the process, and underestimating the natural and physical forces that underpin it. This tension must always be kept in perspective as we continue to remind ourselves that the percipient is embedded in an immanent process of prehension that includes both compression and expansion.

To advance the application of Whitehead's thought to music, Jones makes the important distinction between 'pure music' and 'programme music'. There are also distinctions identified between referrentialists and non-referrentialists, with the latter dividing further into formalists and absolute expressionists.

In pursuing the present line of argument, it is important to understand these distinctions.

Pure music, or absolute music, is not connected in any way to an external referent, not even via a title.

Programme music is music that is based on an extra musical or external source, it is a musical form of expression that interprets ideas that exist in either the real world or other mediated modes of expression – a poem for instance.

Referrentialist, the belief that music refers to meanings that exist outside of itself.

Non-Referrentialist, the belief that music refers only to itself, constitutes its own truth as a pure mode of expression. Non-Referrentialists divide into:

Formalists for whom tonal structures have meanings which are strictly musical and nothing else.

Absolute expressionists, who, while acknowledging the expression of emotions in music, regard those emotions as strictly musical, confined to the musical realm and not evocative of the world beyond music.

> The *Referential expressionists*, on the other hand, claim that musical meanings legitimately refer to the extra-musical world, whether that be ontical or ontological. This theory owes its unpopularity presumably to the implication that somehow one shouldn't listen to music as such at all, but rather daydream of swirling torrents and great vistas of the natural world – anything extra-musical in fact (a common perception among non-musical people). We may not favour this theory, but in the present context we may choose to review it and give at least some credence to it, albeit in a rather unorthodox way. This is because the whole point of our discussion is that of art referring to something else. Bearing Whitehead in mind, E. David Martin's bold compression of Whitehead's thought into a single sentence is useful.

> Music more than any other art forces us to feel causal efficacy, the compulsion of process, the dominating control of the physically given over possibilities throughout the concrescence of an experience. The form of music binds the past and future and present so tightly that as we listen we are thrust out of the ordinary modes of experience, in which time rather than temporality dominates. Ecstatic temporality, the rhythmic unity of past-present-future, is the most essential manifestation of the Being of human beings. (Elfyn Jones, 2000: 49)

When listening to music are you hearing something else? Something other than the music itself? Or is hearing the process itself different from the evocation of an external emotion or object? This is the difference between designative and embodied meaning.

What Martin describes is an unfolding of successive instances of sense data, and while sound and music in particular embody this process they are not alone in doing so. Writing and speaking also unfold in such a linear fashion, where sound is unique, though it is in its assumed abstraction, its non-referrentialist status. This combination of temporal unfolding and abstraction, it is believed, means that Whitehead's process can be directly encountered and experienced in and through music (Elfyn Jones, 2000: 49).

As non-referrentialist and rooted in the mode of causal efficacy as temporal unfolding, pure music never quite reaches the stage of symbolic reference to the extent that it becomes mired in subjective agency. Like all modes of experiencing though there are elements of causal efficacy, presentational immediacy, *and* symbolic reference, the point being made here is that in noise, sound, and music there is at least the potential to access the process as real – the world as it is, as it is embodied.

A linguistic distinction in this regard is made between meaning and abstraction:

> Leonard B. Meyer's differentiation between designative and embodied meaning should be clarified. In language, when a word refers to an object, this is a designative meaning. Embodied meaning occurs when the stimulus and the referent are the same. A note, a phrase, or a section of music has embodied meaning, because it points to and makes us expect another musical (not extra-musical) event. Embodied meanings are the internal relationships of an art form, and in pure music and abstract painting it is the very lack of a designative meaning that distinguishes them respectively from programme music and representational painting. (Elfyn Jones, 2000: 49)

This distinction must extend to the practice of written notation where music, as is the case with other linguistic referent/signifier relationships, is made to adhere to a designative structure. In all such relational exchanges, what is being enacted is a process of compression to one degree or another – the process of ordering an infinite number of possibilities into a linguistic structure that affords the formation of meaning that can be shared and communicated. Abstraction does not in itself however, grant exemption from the process of compression. Music (as perceived aurally) can have designated meaning even when it is not prescriptive as in the case of programme music, assigned to specific contexts or aligned with lyrical accompaniment. Even when, in pure music, no specific meaning is designated, the strictures of tonal patterning and the geometrical rules that underpin it are evident as designated algorithms for the organization of possibilities that is *as* prescriptive, if not more, as any linguistic schema. It may not be in 'pure music' or in music alone then that the Whiteheadian process can be encountered but in noise, sound, *and* music as modes of expansion and compression relative to each other where the fully staged process is revealed.

The overall idea of *modes* of experience as it is being deployed here is derived directly from Whitehead's writings. One of the modes to which it is applied explicitly is *compression*. Compression, we remind ourselves, denotes a process of filtering, prioritizing, regularizing, and organizing as a way of resolving an otherwise bewildering state of overload and confusion. This is both a 'natural' process and a rational one, as well as an aesthetic one. It is, however, also a political one. The efficient organization of knowledge, ethics, cultures, and economies is achieved through the use of what might (very) broadly be called language, or what Whitehead called symbolism. It might also be understood as representation. Whatever the nomenclature it currently needs revisiting.

In *Modes of Thought*, Alfred North Whitehead states:

> The doctrine dominating these lectures is that factors in our experience are 'clear and distinct' in proportion to their variability, provided that they sustain themselves for the moderate period required for importance. The necessities are invariable, and for that reason remain in the background of thought, dimly and vaguely. Thus philosophic truth is to be sought in the presuppositions of language rather than its express statements. (1968: vii)

That which is in the background of thought (the mode of causal efficacy) is noise as presupposition. It cannot be thought in its entirety, but its presence ensures that meaning (signal) can, if only for a moderate period, come to presence and attain a degree of importance insofar as it serves to allay the anxiety faced when confronted with the fullness of the universe. But it is in that fullness, in the noise, that truth resides.

Importantly, Whitehead also distinguishes between notions and systems. He refers to 'notions as they occur naturally in daily life. I am [he says] referring to the generalities which are inherent in literature, in social organization, in the effort towards understanding physical occurrences' (1). He goes on to say: 'The verbal expositions of such data must be trimmed, and dressed, and put in order, during any process of systematization' (2).

This is explicitly the simultaneous interaction between expansion and compression (as trimming). Trimming operates in the service of systematization

Notions, as compressed entities (eternal objects), reproduce exponentially to the extent that they themselves require compression so that paradoxically they are better equipped to access the fullness of the noisy presupposition and to reveal truth.

Whitehead identified in language evidence of an emerging symbolic system. It is a system that produces multitudinous notions, 'cerebral trusts', as functioning normative models.

The totality of the system combines actual experience of actual entities with a conceptual/symbolic model. At present (Whitehead's and our own) we over rely on this linguistic model and its designative function. But how, it might be asked, does this relate to the broader 'linguistic turn'? Rather than the constant undecidability of the text and the tyranny of 'difference' that it supports, and the endless musings of human subjects on themselves and their significance in the universe, Whitehead, in retreating into presuppositions rather than statements, initiates a process philosophy wherein the subject and object merge, and out of which meaning as relatively clear and distinct emerges fleetingly as event. The fleeting nature of the event is itself also relative and vacillates between 'permanence' and transitoriness.

As part of this vacillation the requisite condition for variability is invariable. The identification of a systematic process that is both embodied and designative is what underpins both the complexity and simplicity of the universe. It is the process that facilitates the connectivity of all things and the abstract symbolism that, to varying degrees, sustains it. The process makes sense of itself, producing endless possibilities according to some very basic and fundamental steps. Variability denotes an almost endless set of possibilities/potentialities out of which can emerge notions that are clear and distinct, albeit relatively evanescent – though stabilizing long enough to register themselves as important. In total the variable possibilities constitute noise, which in itself is invariable and necessary to support potentiality.

There are reasons as to why this is significant. Understanding one's relationship to the universe, how it is perceived and how it perceives us, in multiple modes from the organic to the technical and the social, has a profound impact on attitudes, beliefs, and behaviours, both individual and collective. Once it is established that the subject/object dualism can now finally be dispensed with, and that time is multiple and relative, new forms of politics – based on the temporality of noise – can be established. Well, maybe not established, but rather set in motion, initiated, designated as 'important' 'for the moderate period required'. Such a politics would propose an ecology of interacting and overlapping systems, a cosmological ground from which incidental moments/events emerge as transitory but relatively stable and ordered.

This brings us to Michel Serres and his work on the origin of language (where what is invariable is noise and chaos – that which necessitates what is clear and distinct as it emerges into consciousness and sensible cognition). Language is presented here as a phenomenon rooted in noise and sound. Even when it is predominantly visual, it is a phenomenon that echoes the ordering of chaos and the quietening/resolving of noise – from bewilderment to meaning. Sound should not be understood in opposition to the image, but rather as a means of opening a more comprehensive sensory ground where language takes multiple forms. Without leaving Whitehead's universe entirely it is important to consider how his thinking has formed a concrescence with others. In this case Michel Serres.[7] Engaging with Serres allows for some of the ideas discussed thus far to be further clarified. It also allows more specifically the deployment of the temporality of noise as a means of supporting the wider ambition of this text to further challenge prevailing ocularcentrism.

In 'The Origin of Language: Biology, Information Theory and Thermodynamics', Serres sets out a compelling explication of noise and its temporal coalescence into language. He describes beings as organic systems that filter noise to produce order through a complex series of information transfer points. The system is both chaotic and relatively ordered and stable, both a raging torrent of noise and an island of tranquillity – entirely transitory while simultaneously demonstrating a relative degree of what comes to be perceived as permanence. In doing so he repeats many of the themes that were present in Whitehead's thinking described earlier in the text. The organization of the elements that comprise the thermodynamic system is described as enduring across a 'bouquet of time' – as the entanglement of rapid and glacial durations.

[7] This volume does not constitute an explicit comparative study of the work of Whitehead and Serres, though such a study would undoubtedly be welcome and useful. It is worth noting however that very significant resonances are present between their respective oeuvres. This is the case specifically in relation to ideas around chaos, order, and their systematic resolution. These are constants in the work of Serres, and Whitehead addresses them specifically in *Process and Reality* (1969: 100–29). Later in that volume he states: 'In any one epoch there are a definite set of dominant societies in certain ordered interconnections. There is also an admixture of chaotic occasions which cannot be classified as belonging to any society. But, having regard to the enormous extension of any cosmic epoch, we are practically dealing with infinities, so that some method is required, rooted in the nature of the case and not arbitrarily adopted' (237). This method is here understood as compression.

There now exists the apparatus, he proclaims, to conceive of the system and its operation, in tangible and temporal forms, through a 'special listening device'. We have encountered already the idea that sound and music might grant access to a process of meaning making, so understanding how that emerges in Serres's thinking and how this listening device might work is important.

Like Whitehead, Serres is concerned with systems. An organism is a system he tells us, but what a system itself is changes across time, and in accordance with cognate discourses ranging from logico-mathematical, mechanical (relating to statues, fixity, and equilibrium), to thermodynamic (relating to energy and movement). In turn these are, independent of time, constant irrespective of the flow of time, and time bound, orientated towards ultimate entropy and breakdown. We might add to this list digital systems, that in one way or another display all of these stated characteristics. A digital system, as mathematical and largely independent of time, serves to enable fixed and infinitely repeatable models (as Krukowski described in *Ways of Hearing*), and can just as easily be broken down into its constituent parts (bits) and endlessly reconfigured across time.[8] All systems though share closure in common. General equations used to describe systems do not operate well if the systems are open in any way. This is a problem if process and change are accepted as features of a dynamic universe. The need for definitive models, measurements, and calculations is fundamental, as is the need to accommodate change and to recognize that factors external to, or excluded from, any system can potentially have an impact upon it.

The philosophical debate between open and closed systems that has existed ostensibly since Henri Bergson's time of writing still has not been fully resolved. Closing things down, fixing them and imposing 'cerebral trust' as a symbolic function of representation is accompanied by the knowledge of constant and dynamic movement, and these incompatibilities are not easy to resolve. Instinct implores regularity and predictability as a means of sensibly ordering the world, yet the intellect exclaims change, opportunity, and potential – the *what if* of speculation and metaphysics. Rather than being occupied with matter and life and how it was lived (*elan vital*), however, at the foundational and

[8] Digital time is extremely precise and largely invariable. However, the world that it measures is less consistent, and occasionally adjustments (leap seconds) need to be made to bring the two systems back into sync. See https://www.theguardian.com/world/2022/nov/18/do-not-adjust-your-clock-scientists-call-time-on-the-leap-second.

Equally, digital material can decay over time. This can occur in terms of physical decay as it relates to storage devices, but also compatibility decay in terms of hardware and software advances.

systemic level Serres tells us, the debate around systems shifted to concentrate on language and text (as discussed later in relation to *Geometry*), a movement that is attributable to the overestimation of the significance of perception in the mode of presentational immediacy and its synthesis with symbolic reference. There was a related shift in the twentieth century to communications theory and the adoption of aspects of thermodynamics in terms of noise, information, and redundancy – ultimately also used to describe a system of signs.

Serres sought to ameliorate this shift. The living organism, he says, is a thermodynamic system/machine (but a very complex one). It is not in equilibrium 'as thermodynamic stability spells death'. If it comes to rest, its issues resolved, then its energy is depleted, its work is done. It must therefore always exist in a relative state of imbalance. Also, it is now constituted as an open system (something that later developments in thermodynamics can accommodate). Despite the imbalance, the system (organic) is relatively stable and is characterized by a particular type of invariance – neither static or homeostatic. But how might we ask, can something be invariant but not static? The answer is that it changes very little over certain timescales yet never actually stands still. Serres uses the notion of a river and its banks to explain this. 'One always swims in the same river, one never sits on the same bank' (1982: 116).

The river as a system of evaporation clouds and rain is always the same, the banks of the river are on the contrary always eroding. This is akin then to the relationship/difference between sound and image where one (sound) eternally evades capture as it portends towards the future, and in doing so is invariant, whereas the other (image) despite fixing points in time and space, mapped on a grid of reasonable expectancy remains variant – or as was described earlier, sound invokes causal efficacy and image presentational immediacy.

In considering this we are drawn back to A. N. Whitehead's discussion of Cleopatra's needle in *The Concept of Nature*. The event, the needle, is a point in space and time he says, relative to other events and always in motion, but this motion is processed (compressed) to produce and *illusion of permanence* – a static regularity. To illustrate, Whitehead says:

> Cleopatra's Needle is on the Charing Cross Embankment. At first sight we should hardly call this an event. It seems to lack the element of time or transitoriness. But does it? If an angel had made the remark some hundreds of millions of years ago, the earth was not in existence, twenty millions of years ago there was no Thames, eighty years ago there was no Thames Embankment, and when I was a small boy Cleopatra's Needle was not there. And now that it is there, we

> none of us expect it to be eternal. The static timeless element in the relation of Cleopatra's Needle to the Embankment is a pure illusion generated by the fact that for purposes of daily intercourse its emphasis is needless. What it comes to is this: Amidst the structure of events which form the medium within which the daily life of Londoners is passed we know how to identify a certain stream of events which maintain permanence of character, namely the character of being the situations of Cleopatra's Needle. Day by day and hour by hour we can find a certain chunk in the transitory life of nature and of that chunk we say, 'There is Cleopatra's Needle.' If we define the Needle in a sufficiently abstract manner we can say that it never changes. But a physicist who looks on that part of the life of nature as a dance of electrons, will tell you that daily it has lost some molecules and gained others, and even the plain man can see that it gets dirtier and is occasionally washed. Thus the question of change in the Needle is a mere matter of definition. The more abstract your definition, the more permanent the Needle. But whether your Needle change or be permanent, all you mean by stating that it is situated on the Charing Cross Embankment, is that amid the structure of events you know of a certain continuous limited stream of events, such that any chunk of that stream, during any hour, or any day, or any second, has the character of being the situation of Cleopatra's Needle. (1920: 166)

Homeorrhesis, or the steady flow of time, is arrested by abstraction and the establishment of cerebral trust to deduce that the needle 'never changes'. What is perceived in the moment is organized in the mode of presentational immediacy to ensure a degree of order and predictability. In reality, if one stares at Cleopatra's needle for a protracted period then one is, without being conscious of it, witnessing the degradation or change that exists in the relational exchange between the obelisk and the percipient. This can be confusing and logically at odds with the earlier proposition by Stengers that nothing exists outside of visual perception. To be clear, what is meant here is that there does not exist a reality *outside vision* and another reality *inside it*. When staring at Cleopatra's needle one may witness a 'static object', but equally one is always experiencing an object that is changing, but just at a rate this is not immediately perceptible.[9] Either way

[9] Again when we seek definitely to express the relations of events which arise from their spatio-temporal structure, we approximate to simplicity by progressively diminishing the extent (both temporal and spatial) of the events considered. For example, the event which is the life of the chunk of nature which is the Needle during one minute has to the life of nature within a passing barge during the same minute a very complex spatio-temporal relation. But suppose we progressively diminish the time considered to a second, to a hundredth of a second, to a thousandth of a second, and so on. As we pass along such a series we approximate to an ideal simplicity of structural relations of the pairs of events successively considered, which ideal we call the spatial relations of

the same thing, an exchange and processing of data, is being enacted and shared. The relationship is a temporal one and as a process it can be embodied and deployed by artists, philosophers, or scientists, as advanced beings not given to assumption. Rather than being based on assumption or immediate recognition, their perception is active in a dynamic relational manner that draws as much from the mode of causal efficacy (as temporal and musical/sonic) as it does from that of presentational immediacy (visual). If the process is manifest and experienced consciously then the limiting tendency of symbolic reference can be overcome. Ordinary 'language' becomes extraordinary.

Whether looking over a protracted time period, or looking, going away, and coming back, in both cases the process of data exchange and interpretation finds enough commonality to infer no change has occurred. If it is the case that we can, in the right mode, experience the exchange in its fullness, there are moments when we don't, moments where the compression of data leads us to rush to judgement. All that is needed is enough data to proceed, and not the full set. The potentially different ways in which perceived data is processed will impact on percipient response and/or judgement. The organization of simultaneous and contemporary data in the mode of perception as presentational immediacy proceeds directly to the symbolic realm. Both the needle itself and the concept of change are abstracted, and the further from the percipient that this abstraction moves, the more the common-sense judgement of permanence and stasis can be confirmed. Depending on the mode of perception then the needle can be either permanent or transitory, or indeed both at the same time. The difference occurs in the processing of the data, not the exchange, which is always faithful (though prone to error).[10]

Like Cleopatra's needle Serres's organism is both imbalanced and in a state of quasi-equilibrium, relatively stable and fixed but equally always moving through

the Needle to the barge at some instant. Even these relations are too complicated for us, and we consider smaller and smaller bits of the Needle and of the barge. Thus we finally reach the ideal of an event so restricted in its extension as to be without extension in space or extension in time. Such an event is a mere spatial point-flash of instantaneous duration. I call such an ideal event an 'event-particle' (1920: 173).

10 It may though be worth thinking more about scale and duration as factors that affect how reformed subjects are able to process the data they receive. If we can't see the obelisk change before our eyes, is it because we process the pace/duration of change and symbolically recognize it as static (as Whitehead would have us believe), or because an essential element of what a thing is conceals itself from our perceptive capacity (as Harman believes)? What happens if we shift from the visual to the audible?

time irreversibly, 'both a syrrhesis and a diarrhesis' – open like a weaved basket (and therefore also a little bit closed), like an eddy formed from flow, a hesitation in time. It is both movement *through* time and hesitation and resistance *to* time. The organism organizes into what Serres calls a 'bouquet of times' (like noise, sound, music) – it constitutes a multi-temporality insofar as it converts these multiple temporalities of vastly varying duration – sea and island, torrential and always changing; steady, consistent, and almost stable – into lived experience properly organized as an open system, even when it seems very closed indeed.

The human is equally a system of interacting subsystems wherein thermodynamic energy is discharged, and in its discharging generates a cacophony of 'noise' – yet it is rarely heard.

Serres shifts from the universal organism to its local functioning parts – homing in at scale to describe a set of interlinking chemical reactions:

> There exist approximately a thousand different reactions of this kind. But at a given moment, for the complex in operation (living), their number, although probably finite, is incredibly large, in view of the enormous molecular population. For an idea of its dimensions it must be placed on an astronomical scale. From a thermal and information point of view, these movements and transformations necessarily generate background noise. And this noise is certainly tremendous, for the numbers under consideration are gigantic. What prevents us from hearing it? Why is the sound muffled, the factory insulated? (1982b: 117)

Systems naturally quieten themselves to bring a degree of order to chaos, as a particular mode of compression (lossless in this case) that abates the potential chaos of the global totality. This, it seems, is comparable to what happens in Whitehead's perception in the mode of presentational immediacy. Like Whitehead, Serres places a self/subject at the centre of the noise/signal nexus, and the relationship is essential to information theory and therefore to his understanding of noise. It has to be perceived – yet organisms are generally immune to the noise and chaos that surrounds them (filtering out as compression) 'it exists but it goes unperceived', or more accurately unrecognized.

> All of information theory and hence, correctively, of the theory of noise only makes sense in relation to an observer who happens to be linked to them. Who is the observer here? The simplest answer would be to say that for our own organic system we are the observer or observers in question. Thus we should perceive this noise, the noise of a complex to which a receptor is linked. I use 'perceive' in the broad sense that this word had in the classical era. We should hear this

> deafening clamour just as we hear the roar of the sea at the end of the beach. It should deafen us, drown us. Leibniz said the following in his language: the cloud of minor perceptions, external and internal should induce a state of discomfort and dizziness; it should prove intolerable. But save for exceptional instances, we perceive almost nothing of this intense chaos which nonetheless exists and functions, as experiments have demonstrated conclusively. We are submerged to our neck, to our eyes, to our hair, in a furiously raging ocean. We are the voice of this hurricane, this thermal howl, and we do not even know it. It exists but it goes unperceived. The attempt to understand this blindness, this deafness, or, as is often said, this unconsciousness thus seems of value to me. We have eyes in order not to see ourselves, ears in order not to hear ourselves. The observer observes nothing, or almost nothing. (Serres, 1982b: 118)

What is going on here is a system of quietening and filtering, a calming of the tumult, but one that accommodates, understands, and appreciates the noise and chaos as an essential medium, plenum, or ground. Noise is not configured here, or in Serres's universe more generally, as a negative but as an absolute positive. This is important because it makes clear the fact that things that are 'annoying' cannot simply be marginalized and disregarded or irradicated, but instead must be integrated into the system and accommodated, even if they are compressed during processing.

How does this happen, this blinding and silencing, this processing of which we are mostly unaware (unconscious)? The organism functions at successive levels, and the levels link and communicate. 'The level in the interlocking series, manipulates, and generally integrates the information – background noise couple that was given off at the preceding level' (Serres, 1982b: 118).

The exchanges that occur between the levels could, theoretically, be explained by a series of equations Serres states, a mathematical model that could map and make evident these silent and invisible exchanges. Such a model would serve also to make sense of Whitehead's belief that all events (as subsystems in the language of Serres) connect and form exchanges in some capacity, including perception and the all-important formation of the subject.

In his explanation of the functioning of this interlocking system, two types of equations are described with specific reference to information theory. One equation might account for information moving between two stations, another equation might include the two stations and the channel (including noise). The value in terms of ambiguity differs between the two equations and changes depending on the observer's position – either submerged in the communication channel, absorbing ambiguity to create meaning/information, or observing from outside and able to comprehend all elements of the system including stations and channels and resolving noise and ambiguity. This almost perfectly describes the distinction that Whitehead makes between perception in the mode of causal efficacy (submerged inside), and perception in the mode of presentational immediacy (at a distance outside).

It is also pertinent to Maguire's *Ghost in the MP3*. While listening to the original song ambiguity is included in the musical perception, and this supports E. David Martin's belief that music is best understood as residing in the mode of causal efficacy. When the digital process is laid bare and all constituent parts, ambiguity included, are observed from a different perspective, the process itself is perceived as it removes the unwanted 'noise' (in this case, that which cannot be perceived by a statistical average of the population) and compressed into meaning. Hence the compressed and uncompressed interconnect in both natural/organic and technological modes.

Serres also speaks of a 'special listening instrument' that would facilitate access to that which was previously inaccessible. The instrument resides in the consciousness of the self as system. Yet it might equally be a technological device that serves as an impetus to consciousness. The apparatus he says is made up of sensations linked to signals given off and received, but it perceives little or nothing of what we might understand as ground/figure, noise/information. What it does detect though is pleasure and pain, at one level it asks, keep, get rid of? From there it moves to more conscious recognition, but how does the system work?

Serres addresses this comprehensively saying,

> This question only made sense if, at the last level, the most comprehensive of the whole system, the present observer, to whom the noise and information phenomena are linked, had at his disposal or was equipped with a special listening instrument. A point of observation is not sufficient; to observe, one also needs the means to do so. Now the apparatus exists: it is made up of what classical philosophy called internal sensations or what different psychologies have successively described as intropathy, proprioceptive, or coenesthesia, and whose functionings have to be linked to signals given off or received by the vagosympathetic system. The instrument exists and functions. What does it perceive? Nothing, or almost nothing, it seems, of what we recognise at the purely physical level as background noise and information; nothing which resembles, with perhaps certain exceptions, a signal – a figure against

> a ground – isolated from a vague and fluctuating cloud, from a multiple halo humming and buzzing at random. It does, however, perceive the signals that we subsume under the two broad categories of pleasure and pain. It receives them and emits them. It is not meaningless to say that it receives signals that we translate immediately into these two words. Thus everything would take place as if pleasure and pain constituted the final state of a general listening, filtered in turn by the set of successive integrations. The final couple, the only one to be perceived, would, in other words, be the last translation, the last rectification of the original physical couple of information- background noise. Of course, no one can call information fortunate and noise painful, for things are arranged in any number of chiasms. Suffering, at least sometimes, is a set of signals which opens up a path of readaptation or strategies for the repair, for the rebalancing, of the homeorrhesis. Here again, a change in sign appears. There must be noise in pleasure and information in pain. But this is probably something we cannot know or evaluate properly. (Serres, 1982b: 119)

This needs a little more detail adding in relation to its relevance here. Firstly, a point of observation is not sufficient. To observe one needs the means to do so. The first thing to note here is that 'observing' constitutes a kind of listening. The 'special listening instrument' is made up of internal sensations described as intropathy, proprioceptive, or coenesthesia – an embodied sense interaction with the world, a kind of physical sensing and awareness of one's own body and its relationship to external conditions. Once a point of view/perspective, the 'at a distance' visual enaction of perception is adequately appraised of its own physical underpinnings as 'aural', it can begin to account for its role in a universe that in equal measure it impacts and is impacted by.

The successive organic levels function as a language, filtering, and rectifying. To explain this a little further, the term *rectifier* (as a process of *uncompressing*) describes a situation where 'what was once an obstacle to all messages is now reversed and added to the information.' (For example, excess data removed to make the MP3 can be identified and stored – restored and rectified in full audio capacity.) Having perceived Cleopatra's needle and made sense of the scene before us, we may reassess when it is understood that the obelisk itself is constantly transforming. Or we may not. Either way there is potential in the *rectified image*, now infused with the dynamism of sound. Organisms then 'can be described as apparatuses which produce language from noise and information' (Serres, 1982: 120). This process develops through the different levels, from basic chemical interactions to the formation of language – from the noise/information couple through the meaning/obstacles couple, to meaning itself. Within each level the signal is 'compressed' to filter out noise and remove potential obstructions so its message to the next level is clear to proceed. From the perspective of that level a potentially uncompressed mode is adopted. All elements, virtual and actual, are perceived in their complex totality in order that different information may be sought and delivered as required. The preceding level is then rectified in terms of the fullness of its potentiality, significance, and weight, necessary for it to function as specifically as it can and does. These levels correspond to Whitehead's modes and refer us back to Stengers and her assertion that events do not exist outside of perception. Instead, something, a quantum of data, is filtered out or simply missed, but there always remains the option to retrieve it, as Maguire did.

The unconscious, for Serres, constitutes the workings of the lower levels (confined to black boxes) of information production and communication that have not yet formed into a language as we understand it and form the perspective of the integrated level of the observer (the self in relation to others, the observer at the chains end).

Again, Serres describes this in detail:

> It is not a unique black box, but a series of interlocking boxes; and this series is the organism, the body. Each level of information functions as an unconscious for the global level bordering it, as a closed or relatively isolated system in relationship to which the noise-information couple, when it crosses the edge, is reversed and which the subsequent system decodes or deciphers. In each link of the series the question of language is formulated and reformulated by the transformation of the message, the channel, and the noise: by translation. In fact, residual background noise is progressively eliminated: what was supposed to interfere begins constructing; obstacles combine to organise; noise becomes dialect. I imagine this occurs from the depths of the molecular chaos, in which information appears in its spatial simplicity and material forms, throughout the signifying and articulated message through the sequence of rectifiers. What remains unknown and unconscious is, at the chain's furthermost limit, the din of energy transformations: this must be so, for the din is by definition stripped of all meaning, like a set of pure signals or aleatory movements. These packages of chance are filtered, level after level, by the subtle transformer constituted by the organism, and they come crashing at our feet, like the surf at the edge of the beach, in the forms of eros and death. In this sense the traditional view of the unconscious would seem to be the final black box, the clearest box for us since it has its own language in the full sense.

> Beyond it we plunge into the cloud of meaningless signals. Perhaps this box protects us from the deafening gasps of the stochastic; perhaps the box serves to turn them back into symbols. The unconscious is the last black observer of chance. It is an instance of order. It, too, turns destructive randomness into autonomy. (Serres, 1982b: 121)

Comprising multiple integrated levels and mediations, the body is an incredibly complex system that generates language from information and noise. At the chain's end is the observer who utters language, at the beginning it is less certain – deeper and deeper levels of cells and molecules (once more the reversal from presentational immediacy to causal efficacy). Regardless of levels though they are all structured around the information/background noise couple and the chance/programme and entropy/negentropy couples.

In relation to the subject/object couple, and once more in harmony with Whitehead,

> The realms of the subjective and the objective are no longer at odds … The observer as object, the subject as the observed, are affected by a division more stable and potent than their antique separation: they are both order and disorder … There is only one type of knowledge and it is always linked to an observer, an observer submerged in a system or in its proximity. And this observer is structured exactly like what he observes. His position changes only the relationship between noise and information, but he himself never effaces these two stable presences. There is no more separation between the subject, on the one hand, and the object, on the other (an instance of clarity and an instance of shadow). (Serres, 1982b: 123)

Much of this echoes earlier work by Whitehead, and inevitably leads back to his mode of thinking. Like Serres, Whitehead enacted a layering at scale of systems, where what emerges into perception and consciousness as event assumes a degree of *importance*. In relation to any process of compression an understanding of what is and what is not 'important' is essential. Whitehead talked of 'trimming data' and deploying systematic processes to handle thoughts that 'throng into our experience'. But any process of systematic selection, or excluding, already has a problem: 'Today, even logic itself is struggling with the discovery embodied in a formal proof, that every finite set of premises must indicate notions which are excluded from its direct purview. Philosophy can exclude nothing. Thus, it should never start from systematization. Its primary stage can be termed assemblage' (Whitehead, 1968: 2).

This is a very significant problem, and possibly the defining problem being considered here. At its core is the reliance on data formulated at the stage of presentational immediacy (predominantly visual), as ordinary language and received wisdom (as abstraction), that is allowed to dominate without critical reflection. It must be replaced by a sonically inflected kind of logic that can accommodate anomalies as well as change, at all tempos, frequency, dynamics, and pitch, rectifying that which has been omitted through compression. Whitehead assists us greatly with this. His description of process elucidates the constant interaction between compression and expansion. In ordering a bewildering universe, percipients as actual entities or subjects immanent to their environment compress into language the vastness of their experience as symbolic reference. These symbolic references in turn form a nexus of linguistic representations or notions that multiply/expand to the extent that they themselves require compression through a mode of systematization. The process then repeats.[11] Philosophy both systematizes and affords the entertainment of notions of generality – compressed and uncompressed data – big questions and the means of addressing them. This requires making the distinction between *Importance (designated through a process of impression)* and *matter-of-fact (what we find in our midst)*, the particular and the general. Particulars emerge from the general and most often retain a considerable degree of generality. Only with further conscious inspection does significant and particular importance arise. It is necessary to be systematic, but only once the full scope of generality is acknowledged and inhabited. 'We must be systematic; but we should keep our systems open. In other words, we should be sensitive to their limitations. There is always a vague beyond, waiting for penetration in respect to its detail' (Whitehead, 1968: 6).[12]

The process of ordering generalities into particulars cannot, if it is to be useful in terms of the ambitions of this volume, be adequately understood without reference to Deleuze, who well understood its significance. To proceed requires that the confusion between difference and repetition be settled once and for all. Too often, what Whitehead refers to as the designation of importance translates into repetition as resemblance and equivalence, as the ordinary use of language takes hold. For Deleuze only that which is singular and unique, that

[11] What repeats is not the blandness of generality, resemblance, and equivalence, but the unique singularity as particular that serves to undermine the immediate acceptance of ordered norms. This can be read alongside Deleuze's description of generality in *Difference and Repetition* (1).

which cannot be substituted through representation, can repeat. It is this unique repetition as pure difference, that when fully acknowledged draws us away from presentational immediacy back towards causal efficacy.[13] It does so insofar as it accounts for the potential interaction of all actual entities in a relational universe that is in constant flux.

Here again the notion of compression and expansion are encountered – *the world as it is* and the world as it is compressed into comprehension. This is not at all the same as *the world as it is* and *the world as it appears* – as Kant proposed. Rather, it describes the ways in which subjects, in their perceiving, seek to make immediate sense of the circumstances. However, through a modulation of states or modes of perceiving they are able to reassess and include additional data to enhance, rectify, or modify the original judgement. That is not to say that they always avail themselves of this possibility.

In a related manner, Michel Serres describes what he calls the third space, occupied by those named as 'third instructed' to develop the notion of perpetual translation between realms, modes of language, and disciplines. How might an artist for instance translate their ideas to a scientist or a mathematician to a musician? We should seek our instruction neither from science alone, nor the sacred, nor any singular form of understanding, but should rather seek to occupy the spaces of transformation which lie between – neither one nor the other but the 'third space'. Hence Serres gives the name 'third-instructed' (*tiers instruit*) to him or her who is able to give up the comforts of disciplinary specialism and risk putting themselves into perpetual translation (Brown, 2002: 12).

This perpetual translation might be understood not just in terms of a call towards a certain methodology, but as an actual process whereby the relational flow between states/modes of being is at work at all times, creating actualities that are always changing. In order to inhabit such an unstable universe, Serres drew on multiple sources that cut cross time and space, with historical figures and ideas traversing contemporary geographical terrains providing moments of lucidity. As an example, he cites the work of Lucretius and the poetic extrapolation, through the figures of Mars and Venus, of the physical world as the source of contemporary thinking around the notions of complexity and

13 Generality, as expressed through resemblance and equivalence, constitutes a mode of engagement (as compression) that brings the world to order. There are moments, however, when it is insufficient and the need for particulars and singularities is evident. How these two modes/features or subjective interventions should interact and in what ratio is the defining question concerning this volume.

simplicity (expansion and compression). Lucretius positioned Mars as the master of order – presiding over atoms that are consistent and amenable to geometric calculation and the reliable laws of cause and effect. It is a mathematical world where predictions can be made and outcomes are predetermined. However, the reign of Mars is placed under erasure by the presence of Venus who presides over a very different world, a world of chaos and uncertainty where prediction is far less reliable if not impossible. It is though, a world of infinite potential. It is the realm of the possible and of the artist – full of nuanced flourishes and of contemplation of colour and shape at the expense of immediate judgement and perception.

For our purposes Lucretius illustrates the more general point that different modes of experiencing or interacting with the world have a significant impact in terms of either closing that experience down into definable, measurable, and predictable segments, or opening it up to potentiality, uncertainty, and speculation. And these modes are not operable in isolation but interact with one another in variable ratio where both provide something important in terms of knowledge production and distribution. They function as part of an enduring process of movement and attrition in a universe of chaos and noise, where waste and error are constant features. In this context they operate as both unfolding and unstable visceral affect, distinct from, and central to the critical distance of the formal intellect: where there is signal there is noise. The important question is, what do we do with the shifting ratio of signal and noise, communication and miscommunication? Do we prioritize one over the other, culturally, politically, economically, and aesthetically, or do we recognize and embrace their necessary coexistence to inform new modes of engagement? Without reservation the latter is proposed here.

Wittgenstein understood this problem, and it is expressed in his well-known statement, 'What we cannot speak about we must pass over in silence" (2009: 89). For Wittgenstein, information that could not be said to form part of a logical structure of language was superfluous and therefore not well disposed to knowing. It posed a potential danger – of misunderstanding or miscommunication, insofar as it stood ready to be variously interpreted. Hence it had to be silenced and effectively confined to a different dimension, one not well disposed to perception, and where until such time as a common structure can be identified, there is no possibility of communication between this 'colloquial' form of expression and the more formal and evidently meaningful mode. But what is this silent dimension? Can it be thought of differently? Or are we forever bound by Wittgenstein's configuration of it? Paradoxically it might be better understood by invoking the figure of noise. As chaotic, deviant, and unruly, noise feeds back into the ordered system causing miscommunication and disrupting the clarity of logically structured language to the degree where perpetual translation is required, as Serres has pointed out. This however is not necessarily a negative incursion, as noise also constitutes an important part of the overall signal from which sensible meaning is derived. Noise, from this perspective, offers a means of describing the dense plenum in which language subsists and out of which it forms, and reforms. As Cary Wolf states in his introduction to *The Parasite* (2007), noise should be understood 'not as the other or opposite of content, but as content's very fiber' (xiii).

Turning silence into sound and noise in this way reclaims what is lost to Wittgenstein, without losing Wittgenstein himself. He was, after all, correct about many things. The world is formed of repeating patterns that can largely be identified and modelled. Yet it also retains the potential to surprise. While Wittgenstein demonstrates the common structure to great effect, in doing so he silences the colloquial nuances, individual utterances, performances (actual entities that repeat), 'unnecessary' flourishes, and waste products, the surprises that language is also capable of producing. For him such communicative 'excesses' simply get in the way. They are exiled until such time as they can be formally accommodated. For Wittgenstein such an accommodation presupposes the possibility of translation between realms that combine to produce an end-product, the common structure, from the raw material of language. Translating from one language to another – or between systems/modes of language created a dichotomy for him, of formal commonality and superfluous material (waste) which in turn established two distinct spaces, the real and the mystical. 'There are, indeed, things that cannot be put into words. *They make themselves manifest.* They are what is mystical' (2009: 89). They exist but do not form what comes to be understood as the common structure.

For Serres on the other hand there exists no such specifically designated mystical or waste material, no end-product. Instead, the process of translation is always ongoing (though not always without meaning). Waste is not partitioned, but is always present in the process, present and active as noise. In his book *Angels: A Modern Myth*, he makes this point with specific reference to noise and music: 'The messages carried by our voices [he says] are made up of various components: a basis consisting of background noise; then a musicality of sounds accompanied by phonemes, varying according to the language that is being spoken; and finally meaning' (Serres, 1995: 79). He goes on to say, 'Music is so called because it is the sum of all the muses: it adds to every art – no art excels without it. As an acoustic support and a precondition of meaning, music is a constant vibrating presence beneath our dialogs' (84). Noise, music, and meaning should be understood then, as changeable states that bind together to form a single sonic economy that is characterized by virtual potentialities as much as actual trajectories.

The ideas of sound, silence, noise, and music are foremost in terms of the argument being developed here, and Wittgenstein himself drew on them to fortify his position and to present a clear account of how language, mediated in a range of different technological forms, marginalizes (silencing in the process) unnecessary data or information, compressing it in the service of effective communication.

Wittgenstein, in the seminal Tractus Logico-Philosophicus writes: 'A gramophone record, the musical idea, the written notes, and the soundwaves, all stand to one another in the same internal relation of depicting that holds between language and the world' (4.014).[14]

They are all constructed according to a common logical pattern. (Like the two youths in the fairy tale, their two horses, and their lilies. They are all in a certain sense one.)[15]

> There is a general rule by which the musician can obtain the symphony from the score, and which makes it possible to derive the symphony from the groove on the gramophone record, and using the first rule, to derive the score again. That is what constitutes the inner similarity between these things which seem to be constructed in such entirely different ways. And that rule is the law of projection which projects the symphony into the language of musical notation. It is the rule for translating this language into the language of gramophone records. (4.0141)

Music, notation, gramophone, all share in common the compressed meaning/truth/reality that can translate between each medium (add a computer and what happens?). There is a difference here between Wittgenstein's concentration of the symbolic, which is post-linguistic in its relationship to reality, and Kittler's Lacanian notion of the real, which is pre-linguistic as will be discussed presently.

14 As a mode of engagement – does sound form a singularity that unites the subject and the object? https://www.andrejdrapal.com/digital-vs-analogue-quantum-vs-relativity/.

15 This stands in relation to the original and the version expedited from the residual material in Maguire's *Ghost in the MP3*.

mode of engagement that can learn much from these other modes. The drive or compulsion to compress delivers a degree of systematic order but it must never be read as a static device for delivering solutions to problems that are themselves mobile and dynamic. Rather than racing to a finish line that does not exist, humans must learn to negotiate the constancy of a shifting ground, bathe in its noise and rejoice in its potential.

More from Borges

To facilitate a brief interlude, it is worth pausing for a moment and giving further thought to the very significant contribution to the argument being proposed here, made by Borges. In his introduction to *Labyrinths*, James E. Irby writes:

'Perhaps the most striking characteristic of his writings is their extreme intellectual reaction against all the disorder and contingency of immediate reality, their radical insistence on breaking with the given world and postulating another' (15).

The contingency of immediate reality is significant. As Whitehead and Serres have shown, our perceptive powers are variegated between different modes, each striving for clarity and 'peace of mind' through compression *and* through clarification and accommodation of nuance and detail through expansion or extended contemplation. That something else may be possible, or that things may be approached in a different way, are ideas that must be ever present.

> Borges is sceptical as few have ever been about the ultimate value of mere ideas and mere literature. But he has striven to turn this scepticism into an ironic method, to make of disbelief an aesthetic system, in which what matters most is not ideas as such but their resonances and suggestions, the drama of their possibilities and impossibilities, the immobile and lasting quintessence of ideas as it is distilled at the dead centre of their warring contradictions. (16)

It is not the case that 'mere ideas and mere literature' are ultimately futile, but that they take their rightful place in a process, or 'aesthetic system' wherein, as actual entities (symbolic in nature), they interact with other actual entities, emerging as they do from modalities of perception to constitute the immense possibilities inherent in the creative imagination. To write in such a way, that so eloquently and vehemently resists the drive toward representational closure, is to expose the process of compression itself, to expose its limitations and also to celebrate its powers.

Through his writing:

> we are transported into a realm where fact and fiction, the real and the unreal, the whole and the part, the highest and the lowest, are complementary aspects of the same continuous being: a realm where 'any man is all men', where all men who repeat a line of Shakespeare are William Shakespeare. The world is a book and the book is a world, and both are labyrinthine and enclose enigmas designed to be understood and participated in by man. We should note this all-comprising intellectual unity is achieved precisely by the sharpest and most scandalous confrontation of opposites. In 'Avatars of the Tortoise', the paradox of Zeno triumphantly demonstrates the unreality of the visible world, while in 'The Library of Babel' it shows the anguishing impossibility of the narrator's ever reaching the Book of Books. And in 'The Immortal', possibly Borges's most complete narrative, the movements towards and from immortality become one single approximation of universal impersonality. (19)

This says so much – the interoperability of fact and fiction, the monism of continuous being that forms a universe to be participated in by humans. Not a posthuman world then, but a world for reformed subjects, and it brings with it an opportunity for them to get things if not right then a little less wrong.

'his miniature forms are intense realizations of Poe's famous tenets of unity of effect and brevity to the exclusion of 'worldly interests' … Greater and more important than his intellectual ingenuity is Borges's consummate skill as a narrator, his magic in obtaining the most powerful effects with a strict economy of means' (20).

In resisting the tendency to compress that is present in linguistic forms of representation, Borges somewhat paradoxically enacts the services of compression itself insofar as he reduces the linguistic process to miniature but precise instruments of perception, critique, and communication. In contrast the mind of Funes is the chaosmos,[16] the universe of endless possibility, a purgatory from which release is only guaranteed via recourse to the reason and logic that he lacks. That those who have access to reason and logic should be grateful is a difficult proposition to dispute. Yet it somehow feels a little too clinical – too cut and dry, too universal, too (post) modern. Does each individual go willingly into

[16] 'Chaosmos', initially a term coined by James Joyce, but also a book by Philip Kuberski (1994).

a universe of signs, accepting the Platonic contract that there is much that cannot, and maybe should not, be known? What if we do not go willingly though? What if the contract is political and economic rather than simply rational and logical? Would there not always be a curiosity, a compulsion to know whether that which has been discarded or designated as peripheral might be preferable to what was retained (as knowledge)?

Returning to the task at hand, it is now imperative that we attend to the discarded, the superfluous, the unwanted and the different (albeit 'systematically'). Indeed, these are recurring themes at present – ecological catastrophe, national retrenchment, economic crises, anxiety, and war. All of these have become consistent tropes in recent times and are related to the notion of miscommunication, over-communication, and overtly false, or 'fake' communication. They should also be related to the notion of constant uncertainty and unreliability in terms of our engagement with the world, where meaning once arrived at may need to be renegotiated almost immediately – over and over. Rather than moving inexorably in a linear progression, that which we thought we had left behind, returns – it is folded back into the present, and we can witness this everywhere. How do we deal with this? How do we turn silence onto noise, or cast a light across the whole stage without being cast into a catastrophic existential shadow of global/universal proportions?

The superfluous and the inexpressible, or that which has not yet been contained in language, in Wittgenstein's universe, must now be given due attention. They inhabit the same realm from which meaning and logic emerge as that which can be expressed or spoken about, or recognized and made sense of. Such sensible manifestations, however, insofar as they are compressed, too often stand aside from what is designated waste, or beyond sense, surplus to requirement, and in effect silent. These maligned modes of being, if theorized differently, however, can be understood as 'the guardians of the possible', to again quote Serres, whose silent presence is always a danger, or an opportunity, standing ready to permeate the sensible. Borders, such that they exist or are erected, are malleable, and likely to be breached at any moment. Science breached by myth, and myth by science, back and forth across time and space, as Lucretius, and many others return to the fold as our contemporaries. Myth communicates with mathematical patterns, analogue with the digital, sensible speech with noise and waste, to form a protean tide that compression seeks to tame creating passing moments of order. To fully comprehend this process and our place within it we need to change our relationship to time, and space and the 'unthinking, mindlessness of nature' – self-organization and pattern formation.

This of course carries the risk of distinct voices becoming indecipherable as the mix becomes too dense, too busy, too noisy (and this will be addressed more specifically in the next chapter where Serres's assessment of *Geometry* is set out in more detail). It also risks coherent logic being sacrificed to a kind of premodern chaos, and a submission to mysticism. Yet it is a risk that Serres is prepared to take. He is not deterred by the scale of the task. He begins to plot the complex passages that link these diverse voices, realms, disciplines, and domains. They are passages and channels through which communication is mediated and translated, and of course this means there is always the chance of distortion, or miscommunication and misunderstanding.

To fully appreciate this, it is necessary to begin to lay out a plan for new modes of existence and experience that can address what Serres calls the multiple. This is the universe in all its complexity, a universe of dynamic interaction and noisy exchange, where causal efficacy is recuperated as an essential mode of perception.

Over numerous texts and deploying many elements to configure his argument, Michel Serres moves between the multiple and the singular, the different and the universal with a dizzying irregularity. In *Genesis* the multiple, in *Geometry* the universal. All the while his writing elicits a polite disdain for those who have declared their position to be correct. He opposes dogma and fundamentalism at all turns, instead opting for a position of positive uncertainty, complexity, and chaos. This theme is taken up here to explore how compression and expansion, and the foregrounding of sound (presentational immediacy and causal efficacy) operate relative to each other – always in political, economic, aesthetic, and technological modes.

Describing the multiple, Serres writes:

> I have no idea, or am only dimly aware, where its individual sites may be, I've no notion of its points, very little idea of its bearings. I have only the feeblest conception of its internal interactions, the lengthiness and entanglement of its connections and relations, only the vaguest idea of its environment. It invades the space or it fades out, takes a place, either gives it up or creates it, by its essentially unpredictable movement. (Serres, 1995: 5)

The drive to overcome the multiple, to arrest its chaotic nature, has been the task of rational beings across millennia. As it was for Whitehead and as it remains for us, it is a compelling challenge to theorize the shifting ratio of order and chaos, to ensure that neither prevails and to situate all actants as relative within a process that does not pronounce ontological sovereignty anywhere within its dominion. In carrying out the task the drive to compress, name, measure, and fix has come to dominate, leading to what has already been described as a tendency towards ocularcentrism. The effect of this is that we have become fascinated by the 'unit'.

The multiple is compressed to produce a unit, and this happens in different ways. It can either be a process of aggregation or disaggregation (there is no contradiction here as both assume the independence of elements). In the case of the former all existing elements are squeezed together to form a single entity, in the latter elements are separated and those that are deemed unnecessary discarded, producing a cleaner more efficient and defined entity. For Serres, 'the aggregate as such is not a well-formed object; it seems irrational to us' (1995: 3). The dominant inclination therefore is to discard.

The search for unity, for the rational object that can be neatly classified, understood, measured, located, and named is fraught with difficulty, as was the search for the logically perfect language that Wittgenstein sought. The destination is never reached, as units always appear to be composed of ever smaller units. So, we find ourselves in a situation where we are uneasy with the multiple and its clumsy aggregates, but the one is also evasive or unattainable. Our sensory engagement with the world – and the world with us – and its modes prove confusing, and faced with that confusion we are compelled to settle for abstract symbolic representations and the annexation of that which cannot quite be integrated into a 'logical' system.

'Sea, forest, rumour, noise, society, life, works and days, all common multiples; we can hardly say they are objects, yet require a new way of thinking. I'm trying to think the multiple as such, to let it waft along without arresting it through unity, to let it go as it is, at its own pace' (1995: 6).

This sentiment is shared here, and the related question is asked: is the ear better placed to comprehend the diffuse than the eye? At stake is a new kind/mode of thinking and engagement with the world and its sensory audition through the power of listening as a full body experience. 'By the ear, of course, I hear: temple, drum, pavilion, but also my entire body and the whole of my skin' (1995: 7). Yet the notion of audition is not solely an embodied sensory one. It too has been subject to the incessant drive to calm the tumult of the multiple in search of the unit. Noise, sound, and music would eventually succumb to this drive to arrest – to a degree at least.

Notation as language

Organizing sound into a conceptual model that minimizes the amount of information needed to make sense and to form repeatable patterns and processes – this is the function of musical notation. It is a kind of symbolism, like language and geometry and cartography – ultimately an algorithm.

Early forms of representation and signification came in the form *of* sounds, as words, serving to compress the bewildering volume of sense data that bodies receive. But somewhat ironically sound, to a degree, *itself* resists formal codification.

> The air around us is filled with sounds. Some are annoying, some are pleasing, and some provide us with information we need. Noise, music, language – what they have in common is that they happen in real time. The moment we hear them is the moment in which they exist, and as soon as we have heard them, they are gone into what we call the past. (Forrest-Kelly, 2014: 1)

There is a lot going on here: noise as annoying, music as pleasing, language, representation, and perhaps most importantly the idea of what it means for something, in this case a 'sound' to exist. The ontology of sound is problematic. Sound is fleeting and temporally unstable. The challenge for a long time was to fix it in time and space, to tame its wild elusive nature. In a sense measure it. In doing so a strange geometry ensues. In its originary mode, sound serves to bring the visual to order under the regime of the word, whence sound itself is subjected to visual codification in the symbolic form of writing. The intention here is not to recount the emergence of language from oral to visual (written) culture and back again, but to emphasize the extent to which noise, sound, and particularly music, despite having eventually succumbed to the codified practice of (visual) language as compression, can still resist and remind us that there are other modes of experiencing.[17]

Sound is lived, it is live, present, phenomenological in every sense. But it has also been, as briefly discussed earlier, cited as an object (Schaeffer), as a thing detached from its source (acousmatic). Often, as Christoph Cox described, sound is presumed to possess something primal, embodied, and visceral at the expense of a valid materiality. Ontologically its status remains uncertain. It can be understood as inextricably tied to the listening subject, or it can be

For a fuller discussion on acoustic and visual space, see https://www.jstor.org/stable/42575480.

designated as an object in its own right. It can variably operate as pure, and as non-representational, but also as representational, and as uniquely able to bring into presence and consciousness the process and mode of experiencing that Whitehead referred to as causal efficacy. In all its forms sound constitutes a mode of engagement that is significantly contested in terms of its status and identity, so much so that its capture and ordering has proved enduringly problematic, and paradoxically this is its greatest strength.

To our ears the perpetual present is all we have. Once perceived sound travels onto *our* past but may still exist somewhere as its energy endures. It is spatially and temporally dispersed and for this reason it (particularly as 'pure music') resists (for a time at least) conception. Or thought differently, it is the resistance that requires if not demands conception. It must not always escape our grasp. Audible concepts, as abstractions emerging from the modes of causal efficacy and presentational immediacy contain meanings, are ordered in a temporal flow, as an arresting of their energy in memory. Language ultimately, and initially in its oral mode, becomes the means of countering this resistance. When we hear a word, we remember what it means and recognize it almost instantaneously, it is heard in context. As such the word as sound is both a record *and* a transitory event.

'The oldest recording device, and the newest, is the memory; without it, nothing would be possible. Our memory is amazing in that it connects us with the past, it interprets our present, and it helps us predict the future. What it doesn't do is perform those functions for anybody else' (Forrest-Kelly, 2014: 3). Recordings, like memory, allow the past to speak. This is a significant point. Memory needs an external output if its storage capacity is to be utilized to its full potential, taking its place in the complex network of communication channels, both internally and externally. It forms part of the system of thermodynamic exchange that Serres describes, as well as the process, as it emerges for Whitehead, of movement through causal efficacy, presentational immediacy to symbolic code. In its working, memory is aided and becomes present as output through a number of representational forms, both visual and aural. Painting and drawing for instance are also fundamental recording techniques – fixing something that has moved on or changed. Writing equally emerges as a recording device – a device for recording sound. For many years it never occurred to anyone to 'read silently'. 'We don't often think of writing as a device for recording sound, but the difference in sound between two readers is probably no greater than the difference between one of the early mechanical recordings and the sound it represents – they are recognizably different, and recognizably the same' (Forrest-Kelly, 2014: 4). So, the transitory nature of existence, and particularly the elusive nature of sound, eventually begin to be brought under a degree of control as linguistic and symbolic regimes take hold. Such regimes constitute modes of compression, ordering and facilitating channels of communication between actual entities within and between systems. But does anything escape the regime's explanatory powers? What of music?

> We speak of 'reading' music as though it were the same thing as language … But is it not the same thing – a kind of language? The difference, should there be one, seems to be at the level of meaning, and its uncertainty in music (referential or non-referential). Reading words and reading music have a lot in common: the relationship between the symbol and the sound is much like that between a word and a thing. Both are processes of representation, and of reduction to core meaning. Musical notation follows the same principles as language – left to right etc. … But there are also differences. The message contained in formal written language is very specific, whereas in music is not. (Forrest-Kelly, 2014: 3)

What might the implications of this be?[18]

If writing captures the spoken word (the spoken word is a sound that to a significant degree serves as a means of getting the visual world in order) – a representation of a representation, then the capture of sound in-itself, as non-representational, proved more difficult. In Rome 720 as Forrest-Kelly has stated, 'We have the Architecture, we have the art, we have the description, but where is the music?' (Forrest-Kelly, 2014: 5). Musical notation was established long after other linguistic technologies became standardized. As a process of converting sounds into signs, music was the last to conform. Early notation systems were

[18] What of deconstruction and the deferral of meaning that comes with the linguistic turn?

Derrida criticizes logo and phonocentrism in the Western philosophical tradition that excludes the sonorous voice in its deepest craving for unmediated presence. Certainly in some respect, however, Derrida stays closely connected to the philosophers he criticizes. As is the case with most philosophers, Derrida, keeps the silence intact; his ear is mute as well. 'What interests me is *writing* in the voice, the voice as differential vibration', he writes (*Points*: 140, my italics). Han states that Derrida's work does not accommodate the *phenomenal* aspect of the ear, the voice, or sound. The voice to which Derrida addresses his critical remarks has no tone, no body, no volume (cf. Han: 21). His deconstructive practices do not reach the sonorous domain, the domain of music. He seems to not be concerned with sonorous qualities. More important to him is the idea that texts, either spoken or written, are marked by a certain (non-sonorous) tone, or better, a multitude of (non-sonorous) tones: deconstruction makes clear how texts always prove to consist of an infinite amount of tonalities. Derrida's Ear; http://www.deconstruction-in-music.com/outwork/derrida-s-ear/131.

developed for music that was for the most part accompanied by words, indicating not the note but the syllable.

> Writing down music can be complicated. With a given sound, we might want to notate a number of things: How high is it? How loud? How long? What sound quality or instrumental sound does it have? Interestingly, our modern system of notation chooses to privilege some of these (how high and how long) which are built in to how we write the notes and literally marginalise others (how loud, what sound quality), which are indicated in letters and symbols around the edges. (Forrest-Kelly, 2014: 4)

It is this marginalizing that is significant here. We must attend to what happens around the edges. The history of notation is important in terms of what it does and also in terms of what it does not do. 'The choices that were made in the course of its development are the result of a long human history' (Forrest-Kelly, 2014: 4). And that history includes the inclination towards ocularcentrism.

As a historical process of rational and technical ordering, musical notation was also of great interest to Wittgenstein. Notation adapts over time to reflect what is 'important' in terms of what needs to be captured and communicated (and this refers back to Whitehead and his concept of importance, and to 'leaving out' as a means of compression). The success or otherwise of bringing order to chaos by means of representing and of revealing the structure of meaning in its simplest form rests on the generalized acceptance and communication of symbols, and this was what most concerned Wittgenstein. The identification of core, compressed meaning, stripped of all nuanced and superfluous adornment, was the task of reason, and he turned his attention to music, notation and technological methods of recording and reproduction to underline his position.

Revisiting Wittgenstein and his working through of the relationship between gramophone recording and musical notation serves to underpin his explanation and identification of a common structure present in language. It informs the position adopted here insofar as it deals directly with the processes of bringing the apparent unruliness of sound under linguistic jurisdiction. It is also useful in that it can be read as preceding contemporary practices in relation to the algorithmic compression of data.

Susan G. Sterrett unpicks Wittgenstein's references to sound recording, mediation, and representation in her article 'Pictures of Sounds: Wittgenstein on Gramophone Records and the Logic of Depiction' (2005). Sterrett sets out the significant technological advances that Wittgenstein would have been aware of during his lifetime, notably, in the year before his birth (1889), a wax cylinder recording of Brahms playing the piano, later transferred to a gramophone disc.[19] Wittgenstein was born into a world of recording, exactly at a time when liveness and the experience of listening were in the throes of separation. But recordings, as they were in the 1890s, were of a very uncertain quality, and even as Eddison's cylinders gave way to Berliner's superior gramophone discs, there remained the situation where, in a recording, what was being said was recognizable, but who was speaking was less certain. Though the quality would rapidly improve in the ensuing years, to the point of near exact replication, this seemed to be of little significance to Wittgenstein. The invention of recording was interesting for him at a conceptual level rather than a technical one: he was intrigued by the fact that music could be embedded within grooves on a disc from which sound could be reproduced by the motion of a needle, but even more so by the fact that these recorded sounds could be translated into a written score consisting of marks on paper. Ultimately these two modes or models, like all forms of language, would have to possess common features that would allow for the translation of each into the other in order for their common structure to be identified.

The common logical structure present in such models was a kind of compressed signal, where identifiable meaningful communication, whether written language, musical notes, or the gramophone recording, was present always, unaffected by limitation or interference. It demonstrated an economy of expenditure – excluding miscommunication, or noise, or the always present potential for interpretation by the reader/musician, or the mechanical playback related to equipment and the condition of the disc. What was not immediately present and translatable in these forms of representation, however, was effectively silenced, designated as colloquial, or set to the margins of the virtual or the possible not yet manifest, or manifest only in the mystical sense. The wrong note, mistiming, or the scratch were inconsequential. The point for Wittgenstein was not one of accuracy or perfect replication, but the presence of a common structure that could guarantee recognition regardless of the nuances of reproductive quality. A recording of a certain quality could undoubtedly capture more detail than a score generally does, but that was not the point. The point was that the common structure applied only to the minimum amount of information

19 This recording that was later transferred to a gramophone disc was then more recently transferred to an MP3 file, and this is significant in terms of the translation between modes of representation, and particularly in relation to the work of Ryan Maguire.

required to ensure meaningful communication. This minimum amount is what is present underneath the interference and the unnecessary adornment, unaffected by interpretive role of the performer that carries an infinite potential for miscommunication.

In order for a symphony to be extracted from the score, only a minimal amount of information, in the form of marks on paper, needs to be provided. They can be reproduced as sounds that when heard can in reverse be written down/notated, and so on, back and forth. This relationship between elements of depiction and translatability into performance is in effect what constitutes logical structure shared by the depiction and the depicted in equal ratio. Any examples of nuanced interpretation that may exist within the performance exists outside of the logical structure. Once notated the manuscript must be able to be played. Notes must be able to be played and played notes must be able to be notated. They are thus relative to each other – the depicted and the language of depiction coalesce to produce meaning, independent of specific circumstance or condition. Everything outside of the logical structure is effectively 'sent to Coventry'. It is noise that paradoxically must be passed over in silence – not spoken of or to in formal terms. Yet questions remain. What is outside of the logical structure of the symphony, or more accurately, what out of all that is inside is superfluous? Is the 'outside' essentially an affective and distinct mystical realm? Are accidents, mistakes, errors, and miscommunications necessarily designated as mystical? Finally, should the mystical itself be brought under the control of logic or can it be liberated as an energy that is present without demanding representational recognition, and therefore not mystical at all? The position adopted here is that in a chaotic universe, that which is logical is bound to the mystical and that which is mystical is only the unrealized potential of all that ultimately becomes manifest in language. The fate of each, if they are separated, becomes political and dependent on a human vocabulary of expediency, choice, and hierarchy. So, the task is to breach the divide and to recognize the extent of self-organization in a universe where the logical and the mystical are unstable and fluid pattern formations that inhere in each other and emerge from a single material plenum.

The invention of the phonograph that so interested Wittgenstein, in terms of reduction to core meaning, was also of interest to Friedrich Kittler, but in a significantly different way. He has described phonography as a means of capturing the temporal nature of sound in such a way as to render the static signification of formal notation, a relic of the visual age. 'Audio recording registers the messy, asignifying noise of the world that, for Kittler, in a heterod materialist rendering of Lacan, corresponds to "the real" – the perceptible plenitude of matter that obstinately resists the symbolic and imaginary orders. "The real" Kittler concludes "has the status of phonography" (p. 16)' (2010: 154). This 'sonic real' resists the tendency of representational media to compress information and instead unleashes 'what Kittler (1999[1986]) calls "the waste or residue that neither the mirror of the imaginary nor the grid of the symbolic can catch: the physiological accidents and stochastic order of bodies" (p. 16)' (Cox, 2011 155). Or what I would term noise. This is not only true from a technological perspective, but also in terms of philosophy, aesthetics, and political economy, or 'sonic economy' where that term describes a chaotic system that can account for the world in its expanded mode without being bound by representational certainty or rigid equivalence. The word *economy* on its own describes a mode of compression where measures, rules, and static practices are instantiated to serve reliable and predictable illusions of the world as picture for the purpose of exchange. The addition of the *sonic* recuperates that which is lost to the ocular economic models, to reveal truth as an emerging process of unstable, or at best partially stable, relations.

As Kittler accounts for the impact of phonographic technologies and their correspondence to the real, we must extend this to include digital processes/systems, through the MP3 that has already been discussed, up to more recent developments in artificial intelligence (AI). As language (as symbol) enters the digital domain, it is important that it be understood within the linguistic framework that has been set out thus far. We are it seems at a point in time where the very structure of digital code, as a linguistic form in its own right, is merging with or returning to its symbolic origin in relation to the word and the voice as it orders a chaotic universe, making it comply with an abstract representational logic that 'does our thinking for us'. If this is the case, then there is an even greater need to develop (open) systems that are prefaced upon an enhanced sonic sensibility.

The intention here is not to engage in a protracted discussion of AI, but simply to recognize the digital as an important extension of the linguistic mode of compression under consideration within the broader context of this volume, where the emergence of symbolism as a mode of compression is understood not as a dualistic subject/object division but as a natural monistic process, even in its technological capacity. The entanglement of organic and technological modes of perception and communication are evident in current human/machine interactions.

AI *prompt engineering* is a technique of framing instructions, both written but increasingly spoken, in terms most clearly understood *by* the system, so it returns the results that most closely match expectations. It describes a practice of communication between systems and as such is akin to the process described by Serres where the instruction or request is compressed for ease of communication, but where the data base, while responding in an equally compressed manner, reserves the uncompressed potential in relation to future communications. A universe of almost limitless possibilities is ordered according to a sign regime that is multiple rather than unitary.[20]

In terms of *image generation* AI relies on the assembly and analysis of countless tagged images. The tag, a word linked to an image, recalls the designative feature of linguistics discussed previously. It means images come with some kind of description of their content already attached. These images and descriptions are then processed through neural networks that learn to associate particular and deeply nuanced qualities of the image – shapes, colours, compositions – with certain words and phrases. And this can be identified as a kind of perception in the mode of presentational immediacy. These qualities are then layered, as they are in human processes of perception, on top of one another to produce new arrangements of shape, colour, and composition, based on the billions of differently weighted associations produced by a simple prompt – as a very significant act of compression.

Such an act requires very sophisticated mathematics to explain/account for it, and in the digital present such capability is entirely real. Advanced algebra, calculus, complex geometry, and algorithmic ordering are able to describe a universe in a manner that while including the contribution of a human subject, can just as easily comprehend a universe without humans, to the delight of the speculative realists.[21]

As previously referenced, Whitehead saw mathematics (and here the digital is included in this) as a more advanced symbolic register, a means of codifying reality that did not draw on human rationality, so its encroachment into the 'subjective realm' of rational linguistic symbolism, both visual and aural, is significant. Through developments such as prompt engineering, image

[20] The total (singular) dataset as sought by AI is an exercise in linguistic totalitarianism – doomed to fail at the altar of meaning and representation. The pursuit of a theory of everything will always be prefaced upon the paradox that what will be discovered is the fact that chaos and unpredictability are at the core of knowledge and not its nemesis.

[21] This applies most explicitly to Alain Badiou, but also to Quentin Meillassoux. For an outline of Badiou's position see https://www.artforum.com/features/being-by-numbers-202925/.

recognition, and the more ubiquitous no code future, 'artificial intelligence' appears to be following an uncannily similar trajectory to that systematically captured by Michel Serres, in his *Origin of Language* systems, and we remind ourselves that what a system is changes over time and now includes the digital, that succeeds not by excluding the existing intelligent processes (in this case humans) but incorporating them. Perhaps this is what scares us so much.

'No code future', then foresees a time where the most efficient way of interacting with intelligent systems would be through the medium of human language. Not so much a no code future then as an old code future. The digital turn and related AI debates are part of the wider linguistic system of information processing that has been set out here. Currently language in its digital mode tends towards the ordinary and to presentational immediacy rather than embracing the extraordinary potential of causal efficacy and noise. Linguistic systems in all their forms operate to deliver an element of order, quietening, or silencing noise as they go, but always themselves operating within the noise without which they could not function. Each in its quietening and in its offering of moments of reflective calm, contributes in increments to a restaging of noise, to be resolved anew.

Concluding remarks on this section

Compression is a feature of political economy, philosophy, art, music, and the digital, and these activities or disciplines combine to form a set of complex interactions that when properly configured as systems, subsystems, and processes, can account for the symbolic representation of the real without recourse to a sovereign subject upon whom that real depends. Compression generates residue which is often followed by the desire to recuperate. Has there ever been a standardized method or language that was not contested – tonal/atonal, harmony/noise, where the stringent formulas of production were not pressured to include the excluded? The answer is no. But these states are not taken in turns in a dialectical fashion but rather exist as aggregates of elements that ebb and flow, coexisting as singularities, sets, units, in a sea of multiplicity, where the drive to contain is always in vain. No sooner does Wittgenstein settle arguments as new ones emerge to require recalculation (even though it took him some time to realize this).[22] Whereas Wittgenstein dispatched the unknown

22 Wittgenstein's later works understood that his earlier claims to have resolved all that philosophy was capable of had been made in haste.

into a realm of silence, Serres celebrates it as noise, the all-important medium through which all reason and creativity flow. And it is in noise that this volume resides.

> Background noise is the ground of our perception, absolutely uninterrupted, it is our perennial sustenance, the element of the software of all our logic. It is the residue and the cesspool of our messages. No life without heat, no matter, neither; no warmth without air, no logos without noise, either. Noise is the basic element of the software of all our logic, or it is to the logos what matter used to be to form. Noise is the background of information, the material of that form. (Serres, 1995: 7)

Serres's use of the concept of noise, as we have seen, is complex. It describes relations, passages, movement, potentiality, but also excess and chaos. This is why his thinking is so relevant here. The contemporary digital environment with all its connectedness and potential produces a surfeit of information. When the excess of language reaches a critical point in terms of volume it returns to noise as the energetic site of potential and movement, where the calm becomes the storm – like a picture that contains so much data that it contains no information at all (white noise/black noise).[23] As such the virtual should be understood not as an alternative realm after all but just another fleeting pattern in the swirling sea of unpredictability that is the multiple – a map of the real so significant in scale that it is the real – not an algorithmic mirror *of* the real, because the real itself is an algorithm – not an algorithmic representation of an algorithm – indeed not standing in relation to the real at all – just the real itself being itself, compressing and expanding incessantly.

Compression is mathematical, linguistic, natural, *and* synthetic, sometimes musical, and always political. The figure of compression as it is developed here, though derived from studies in sound and acoustics, has much wider implications. Compression serves to bring a degree of order and enhanced comprehension as a qualitative and quantitative feature of the reception of information. Yet as the imposition of order over chaos it is not always and inevitably beneficial. Stripping something, an entity, object, concept, equation, or composition, down to its most simple form may have certain clear benefits; it may mean things are easier to work with, more able to be comprehended, organized, or stored. But as Serres reminds us, it is a process that is flawed when we assert the belief that

[23] As in Balzac's *Gillette or The Unknown Masterpiece*.

energy and information should not be separated. Language in all its complexity emerges from, and is best understood within, the context of an organic system, where energy and information are indivisible (Serres, 1982: 121). Language (including the digital) has become too readily subject to a mode of compression where its 'noisiness' is forgotten. Wittgenstein expedited this forgetting when he explored the superfluous nature of language and the related errors of philosophy that needed be remedied through logic and a renewed understanding of symbolism. In reducing meaning to its core Wittgenstein at least recognized the presence of an excess that must be moderated. And this it turns out, despite clear and obvious differences, is not so different from the position that Whitehead advanced insofar as it echoes his inclination towards presupposition and structure as opposed to statements. Equally, and somewhat ironically, it is also not so different from the work of Michel Serres himself who is clear that while it is noise that sustains it, language, when it becomes excessive, is also something from which the soul needs salvation.

What does Serres mean though when he talks of salvation from language, and how is salvation achieved? In basic terms he is referring a state of information overload, and the sensory bombardment of potential interpretations that were in part unleashed by advocates of the linguistic turn and the postmodern milieu. Salvation from it is achieved via recourse to the imposition of a systematic mode of representation, as it was at the time when language first began to emerge, reminding ourselves as we must that any such system should ideally be open (which too often it is not). The idea of salvation is of course problematic, particularly when deployed by advocates of a variety of versions of political closure, but it is understood here as describing the simultaneous processes of compression and expansion that together comprise an entire system of interaction and exchange. Through a compressed logical engagement with language on a journey into the expansive noisy sea of the multiple, it is recursive, hesitant, and decisive in equal measure. It is a system that is organic, technical, aesthetic, economic, and perhaps most significantly political. And it is to this political dimension that we must now attend as this section moves towards its conclusion.

Language comes to form patterns and rules as Brassier has demonstrated, and in doing so it becomes embroiled in what Jaques Ranciere calls disagreement. It is a disagreement about the validity and distribution of sense data, a form of compression that finds its home in the exchanges of 'ordinary language', in error and in judgement, as salvation from the torrential potentiality of the universe.

> According to his definition, *disagreement* is neither a misunderstanding nor a general lack of comprehension. It is a conflict over what is meant by 'to speak' and over the very distribution of the sensible that delimits the horizons of the sayable and determines the relationship between seeing, hearing, doing, making, and thinking. In other words, disagreement is less a clash between heterogeneous phrase regimens or genres of discourse than a conflict between a given distribution of the sensible and what remains outside it. (Ranciere, 2003: xiv)

Ranciere also introduces the concept of the 'police' or 'police order' (xiii). Gabriel Rockhill defines the concept in terms of a division between those who participate and those who are excluded – yet another form of compression. This is compression in its political mode, or politics as itself a mode of compression.

> The *police,* to begin with, is defined as an organisational system of coordinates that establish a distribution of the sensible. This law implicitly separates those who take part from those who are excluded, and it therefore presupposes a prior aesthetic division between the visible and the invisible, the audible and the inaudible ... The essence of *politics* consists in interrupting the distribution of the sensible ... Those who have no name, who remain invisible and inaudible, can only penetrate the police order via a mode of subjectivation that transforms the aesthetic coordinates of the community ... Democratic emancipation is a random process that redistributes the system of sensible coordinates without being able to guarantee the absolute elimination of social inequalities inherent in the police order. (xiii)

This requires that we accept that the sensible can be appropriately manipulated in or out of our preceptory range – through processes, not just as linguistic phantasms but as real, and as 'law', de facto and de jure. The 'prior aesthetic division' assumes that the visible and invisible, the audible/inaudible exist as a controllable ratio. It is not the case, however, that the invisible and inaudible do not exist but that they assume the status of dark matter with weight and potential. What the relationship is between what is retained and what is lost in compression is crucial, for as Ranciere points out democratic emancipation is a random process of redistribution without guarantee of success. It requires the intervention of a subject that emerges (through subjectivation as an acquiring of agency that enables the pattern to be reformed by improvising in terms of rules) to alter the existing sensible landscape. All of this is underpinned by an 'organisational system of coordinates' that serve to select, separate, and exclude, describing almost perfectly Maguire's *Ghost in the MP3* as a distribution of the sensible in a digital context. But how, if at all, can the pieces be put back together? This is an organic, technical, as well as a political question.

What happens when individuals, groups, and the potential of what they have said or might say is silenced, marginalized? It continues to exist, albeit in a peripheral sense, because the distribution of the sensible is also a filtering of the sensible that in its contemporary guise occurs as an onto-technological phenomenon where more than ever access to specific modes of representation is determined by the distribution of resources across the global political economy.

Language as compression is inclined towards definition and certainty. It imposes order on a chaotic world. Systems of language, as they form lexicons and structures, likewise impose order, and set reliable expectations and forms of meaning and exchange as an 'economy'. This order, stability and reliability, is however always temporary (but none the less real). It is exposed to a host of pressures, external, environmental, but also internal in relation to signal and noise. Often language is well equipped to adapt to such pressures, showing itself to be less defined and fixed than might previously have been thought (as argued for by proponents of the linguistic turn). At times however its hold on reality is broken and it becomes meaningless, or at least unstable, while at other times it is able to hold firm.

Such relative stability and adaptability (as the bouquet of time), when faced with pressures and challenges, is also present in political economy, where systems and modes of organization redefine themselves constantly. Yet something is different when it comes to political economy, particularly in relation to the subsystem of capitalism whose advocates are in denial. They steadfastly cling to the pattern while arbitrarily changing the rules. But it cannot hold forever, the 'island' is eroding. Capitalism is in crisis – a systemic and linguistic crisis where definition and meaning have been stretched to such a degree as to render them almost incoherent. The lexicon as ordered set of rules has been compromised to such an extent that any original meaning has been lost. In an attempt to conserve the system, its most committed protectors have breached it. This has led in some areas to a critical state of polarization (and the restaging of dialectical conflicts) as a deliberate means of defence as last resort. In the most 'advanced' cases capitalism, even where it is at its most 'liberal', is constituted as an idealized monument to human achievement, a system with structures and rules ensuring that the incremental generation of wealth (numerically measured as a zero-sum equation and presented as compressed data) is distributed in the best possible way'. This is only true, however, if a degree of collateral damage (waste) is accepted. It is not accepted here. The audible signal that endures through noise is being enveloped as the system rectifies and the data begins to decompress. Paradoxically perhaps, the challenge to the validity of advanced capitalism relies on the idea of data as an expanded system of communication and representation that includes everybody and everything, where all anomalies are part of the system in its entirety.[24] Such a system becomes increasingly viable if the digital is understood, as it now must be, as a complex, noisy process, and not as a definitive and closed linguistic structure.

The intention here has never been to repeat the postmodern mantra of difference (discussed further in the final chapter) where meaning and truth are eternally deferred. Rather the aim has been to accommodate compression as a naturally occurring facet of an expanded system that is an assemblage and prefaced on the idea of the multiple rather than the unitary. We must always remind ourselves to keep our system open without being resistant to specific logical solutions (in the modes of mathematics and geometry for instance) where they work. This requires organic, natural systems (including humans) becoming more closely imbricated with synthetic, scientific, and mathematical (and ultimately digital) systems.

Humans (to a certain extent) are themselves subject to a process of compression, reducing them to a series of mathematical datasets that co-mingle with the object world rather than separating from it as is commonly proposed in classical models of Western philosophy.[25] Such classical interpretations, in that they are accompanied by interminable and relentless identity politics and the insufferable 'creative' and 'artistic' reflections on the 'self', require significant adjustment. More recently human compression, through genome mapping, spectral measurement, stratification of parts, personality types, behavioural taxonomies, and so on have been framed in such a way that can serve to challenge this historical separation between subject and object. To be human is to be present, not separate, and party to a set of complex chain reactions, input data and advanced processing/processes – from causal efficacy, recognition, and

[24] The 'rules' and patterns that constitute the capitalist system it seems can be improvised almost at will to protect the interests of its exponents. So much so in fact that the very principles it advocates can be seen as arbitrary and open to challenge. As a system that excludes, it presents the blueprint for its own demise.

[25] Foucault deploys the notion of the double, empirical transcendental (being as both an organic entity like all other entities and a sentient thoughtful being that processes information and sense data to produce knowledge and meaning) https://www.cambridge.org/core/books/abs/cambridge-foucault-lexicon/double/C24A5C818958624ADE91F0F7B5F1CAA1.

familiarity to symbolic reference and the production of concepts. The perception of colour, for example, as Whitehead claimed, is perception of something real that exists externally – one sees the colour that the thing is. This causes either a direct response/reaction/tropism or a familiar recognition that emerges as symbolic reference that can then be processed into a concept – a form of compression that develops to order an otherwise unruly universe. Humans now take their place in the object world, but paradoxically they do so as reformed creative subjects (as objects) possessed of the systematic and functional ability to affect change.

Overestimating the significance of presentational immediacy, and the status of the subject that it produces, has served to underpin a tendency towards compression, as a mode of ordering chaos into reliable and static monuments of knowledge. However, this leaves much that is unknown, unconsidered, unimportant. Alternatively, by not underestimating the significance of causal efficacy in a universe of connectivity, as Whitehead suggests, the production of new knowledge can be accommodated. Not only compression but also expansion is accounted for, in shifting ratio, as facets of a systematic process of emergence and renewal. The reformed subject as embodied sonic being is then empowered to rearticulate the universe as dynamic system of energy exchange. Once the subject/object dualism, in its traditional sense, is dispensed with, humans can be allowed to re-enter the arena, newly appraised with the knowledge that their being, perception, conceptions, and actions necessarily coexist with other universal elements as figures emerging from ground or noise. This shifting of position constitutes a new paradigm wherein humans take their place within an ontological space that depends on them no more or no less than they on it. The reformed subject can then speculate once more.[26]

For Whitehead, and of course for many others, the speculative act was equally a creative one, and artists could be taught to harness its power. Inherited from the potential for error, it is a process of contemplation. As distinct from the simple and immediate act of recognition – a mode of compression where all 'unnecessary' sense data is dispensed with or arranged into contemporaneous patterns – for the artist sense data is *expanded*. It's easy not to be an artist, to be one requires training. Art is a mode of expansion, and noise, sound, and

[26] If all forms of speculation, or indeed all philosophy, involve compression as part of the drive to 'make sense' then a political consequence of this is the imposition of one mode of thought at the expense of all the other possibilities. Maybe then, we need to turn to 'non-philosophy'. To do so exceeds the ambition of this volume but non-philosophy is a very important contribution that needs to be explored. See Laruelle. 2017. *Principles of Non-Philosophy*.

music are embedded in art when it is understood as a process that does not succumb to immediate perception of sense data but instead favours prolonged (temporal) contemplation and revealing. This forms the basis of a critique of compression and efficiency as guiding principles of reason and logic that draws upon the temporality of noise as an enduring mode of engagement. Once again, compression and expansion do not form a simple dualism but are inextricably linked to language in all its forms, its warp and its weft – they are distinguishable but not separable.[27]

So, symbolism beyond simple recognition is a deeper act, practised by advanced organisms. Perception of colour, shade, and shape are symbolic of some other register/dimension of our experience – beyond simple recognition, they are qualitive facets of a transitory reality that moves at variable rates. The reluctance, on the part of the artist, to jump straight from recognition to symbol, and in the process instantiate 'ordinary' language use, constitutes an aesthetic mode wherein the extraordinary becomes possible – not as an abstract or subjective judgement but as a real and immanent expression of potentiality.

This is important. The relationship between advanced organisms and 'high-grade' acts of recognition lies at the heart of the interaction between philosophy, science, and art – a coming together that is most alive and energetic in noise/sound/music, as central components of all art. If, for Whitehead, artists were trained higher beings who had acquired the power of contemplation, they have over time come to limit the extent of their contribution by prioritizing the visual. To counter this, scientific, mathematical, and sonic modes of engagement have been prioritized as a means of augmenting and moving beyond the visual register. A sonic sensibility further informs a mode of engagement that can reveal our relationship with the real. Heidegger showed us a system of prolonged temporal contemplation and questioning that despite revealing a kind of truth, relied

[27] Our task is 'to bring language as language to language' (398). This task is like a formula. The word 'language' is used three times – yet always differently (cf., 418) – but what coheres these three usages into one is the selfsame that conjoins that which is held apart: 'the formula directs us to a weft of relations in which we ourselves are already interwoven. Our proposed way to language is woven into a speaking that would like to liberate it in order to present it, giving utterance to it as something represented – which straightaway testifies to the fact that language itself has woven us into its speaking' (398).

Weft: (pl. woof) the horizontal threads interlaced through vertical ones (warp) in cloth. 'Warp and woof' designates the foundation upon which things are built woven.

https://aquestionofexistence.com/Aquestionofexistence/Heidegger/Entries/2014/11/17_Heideggers_On_The_Way_to_Language.html (accessed 7 July 2024).

in the final assessment on human rationality and as a result fell back into the correlationist trap.[28] Whitehead did not succumb to this fate and has provided us with a template for a process that, although accounting for the role of humans as significant, does not rely on them to validate reality. The speculative realists, and Harman in particular, have attempted to go a step further in substantiating the real, and in the case of the most ardent declarations of realism, describe a universe that while real in every sense, is ultimately withdrawn, private, and obscured (from view) from humans and their epistemological frameworks. They describe a reality that might only be understood via recourse to mathematical models rather than preceptory ones. Whitehead of course had already alerted us to the symbolic significance of such a mathematical mode, and here it is accompanied by music as a means of countering the withdrawal from perception of the objectively real.

In his introduction to *A Thousand Plateaus*, Brian Massumi says: 'On a strictly formal level, it is mathematics and music that create the smoothest of the smooth spaces. In fact, Deleuze and Guattari would probably be more inclined to call philosophy music with content than music a rarefied form of philosophy' (xiv).

Philosophy is then a form of music, but music itself cannot do the work of philosophy. So, music becomes a model *for* philosophy – for doing philosophy musically, as a temporal unfolding with very specific mathematics and geometries underpinning it. This is taken to be largely unproblematic here, though there is, as has been laid out in previous sections also, an inclination towards recognizing in music something of a 'rarefied form of philosophy'. This is particularly the case in relation to temporality, duration, and process. We have already seen how Whitehead can be given a musical inflection, and how Brassier enabled us to think of improvisation, in a musical sense, as a philosophical process. As the process unfolds, noise, symbolism, language (spoken and written), and music

[28] How do Whitehead and Heidegger compare in this regard? One key question here is how the notion of abstraction is understood – whether recognition proceeds immediately to symbol as an abstraction, or whether sustained questioning prevails. To a degree reading Heidegger vindicates Whitehead and his articulation of sustained and contemplative engagement, even if in the final assessment, the subject is not sanctified as the sole arbiter of Being. Heidegger as philosopher was also an 'artist' in his use of words and language, expanding further and further, exhausting all avenues of investigation until meaning emerged. His writing is in a sense musical – a patterned mode of improvisation. Whitehead on the other hand, as a mathematician, sought a kind of algorithmic solution wherein process unfolds perpetually. Again, as is generally the case with the argument being developed here, the two are not mutually exclusive.

form a historical continuity where noise is an originary state, and with music the last to be formally codified, if not always coherently organized.

Music can be understood as a symbol but is also exposed in itself to the symbolic realm of representation and perception. This is what makes it so interesting. The difficulties inherent in assigning to music representational qualities alongside the difficulties in subjecting it to an effective representational regime of signs make it particularly suitable as a model to be applied to complex material environments. Symbols come in an array of shapes, sizes, pitches, and timbres. History shows them to be fragile and subject to change – sometimes the habitual response is ignored, like in Brassier's text where the learning of rules, so that they may be ignored, constitutes a mode of creative expression understood (in a musical context) as free improvisation.

Moving backwards through symbolic language, music, and ultimately to noise, the chaotic universe has increasingly been brought to some kind of order. As recording devices, forms of representation have disrupted temporal continuity and allowed us to return to the past, and in doing so we are appraised with the realization that we are constituents of a singularity that forms patterns and rules with spatio-temporal phases in relative concert.

Despite retaining non-representational features, music itself has become increasingly encoded by technologies of representation. These technologies have changed what music is and how it is perceived to some degree. The point here is not to castigate formulaic or captured music or its principles of organization. They work to great effect and despite their formulaic nature still provide for rich fields of creativity and artistic expression (like geometry, as will be shown in the following section, not all organizational models are bad). The point is that to only engage with organized modes of perception is limiting – pleasurable maybe but limiting nonetheless. There is a very significant difference between sound as perceived and sound as recorded (including in the form of notation) in terms of time. Recorded sound is fixed whereas perceived sound is transitory, always escaping our grasp. When listening to music, sound is temporally unstable, but as selective combination of possible elements, fleeting convergences, and intensities that include both resonant and dissonant successive events, it is also possessed of a relative stability. Noise on the other hand is the simultaneous presence of everything, undifferentiated, pure potential., and as discussed previously is where truth resides.

If this is the case it is important to gauge the status of sound art, and of noise, as modes of creative practice. To what extent are they themselves forms and expressions of ordering, even when their intention is to bring into experience the full extent and significance of chaos? Any attempt to capture the wild and ephemeral serves ultimately to assign a relative degree of stasis, closure, and predictability, and this is always at the expense of celebrating movement and change. Is the non-representational aspect of sound art (and some music) enough then to reveal processes otherwise hidden? The answer is no, and nor it need be. The truth is not in itself a stable thing, not a fixed point that can be identified and navigated towards. Rather it is a constantly shifting set of relations that reveal themselves to be present in patterns and rules, and the subjective recognition and/or adherence to such patterns and rules is both part of and resistant to patterns and rules. It can never be the case that the pure mode of causal efficacy be revealed, any more than the entirely smooth space that Boulez and Deleuze and Guattari describe can. What noise, sound, and music do instead, in all their relational modes, is to allow us to orientate and disorientate ourselves in shifting ratio between order and chaos, and it is the balance of that ratio that forms our political, economic, and aesthetic realities.

To inhabit a universe where everything is present may be a challenge too great to bear, as Funes found out. The crafting of noise into music and into language, which ultimately also includes music as notation, must necessarily compress and in doing so produce waste. Not all of noise is present in music and language. Noise, music, and language (notation) are ordered to indicate a progressive, and reversible, trajectory of rationality and meaning – moving from one mode to the other in turn through a process of compression, a process of discarding the superfluous until universal comprehension is achieved. Many linguistic modes collide here as they did for Whitehead – the language of architecture, of art, science, philosophy, mathematics, geometry, and music – but always language is where sound and image combine. Noise, music, and language exist simultaneously and in infinite potential ratios, and where/when one state/mode prevails, it only ever has a tenuous hold on permanence at best. Each state is always potentially present in the others – each makes the others possible – not as cause and effect but through modal immanence.

Finally, compression has been posited here as a process of 'filtering out'. It is a complex process that occurs instinctively and unconsciously for most people. It can serve as a positive, in terms of relieving the overwhelming and bewildering aspect of being faced with 'everything all at once', but it can also be a negative insofar as it generates waste and residue which in turn translate into political and economic debris (as our mode/technologies of language and communication impact on other spheres). To counter this danger the 'natural' operation of sense perception must at times be augmented by an acknowledgement of what is not immediately present. This can be achieved through science, but also, as Whitehead shows, through art. The task is to bring both, as embodied in noise, sound, and music, into the political realm.

3

A geometry of listening

This final chapter approaches perception as a process of revealing – one that recalls causal efficacy and adds it to the mode of presentational immediacy by foregrounding sound as an integral aspect of our engagement as subjects with the world. The reappraisal of sound and its significance within the perceptual process is both specific, when it is added to vision to enhance experience, and implied when all aspects of perception, including vision, especially vision, are understood to possess a sonic dimension. Geometry, cartography, science, maths, art, and music combine with noise and logic to form a process where sound accompanies vision every step of the way.

This 'process' derived from the previous analysis of Whitehead is a 'sonic' one, though it is infused equally into other sensory modes. As such it constitutes an audio-visual model as a cartography of fluid points (vectors rather than points fixed on an axis) that are capable of operating within a temporal, transient framework and not just a static one – revealing things that the eye alone cannot register. The eye has instantiated itself within technological spatial mechanisms that have eradicated, by reducing it to representational code, the nuances of space, place, and distance. Hence the need to revisit the practice of cartography, as will be done in due course.

To fully understand the process, it is necessary to return once more to the thinking of Michel Serres. Like Whitehead, Serres draws on the history of mathematics, or more specifically geometry, assigning it a special status, and in doing so offers a glimmer of hope as we wade through the philosophical quagmire of postmodernism, the linguistic turn, and more recently calls for an object oriented speculative realism. A detailed reading of Serres's position in relation to postmodernism in particular serves to underpin the call for a geometry of listening that can mitigate the failings of difference without recourse to an all-encompassing universalism.

In *Geometry* Serres draws our attention to the absurdity of 'difference' insofar as its ubiquity establishes an ideological model that cannot be deviated from, saying, 'But by some perverse paradox, difference ends up imposing itself in turn as a universal dogma that everywhere and always forbids speaking forever and everywhere. Is it only the local that can be expressed globally? This Law without justice forgets geometry' (Serres, 2017: x).

He also points to the extent to which, while the academic community congratulated itself on its variegated gaze and the universal imposition of difference, digital protocols were being installed that would shape, for generations to come, the political, intellectual, and aesthetic landscapes. While the social sciences were busy with their drive towards multiplicity, the hard sciences were building computer networks that would later become so significant as we now know. All the while the humanities were ignored. That was a mistake then as it is now, yet the call for a genuine and meaningful interdisciplinarity grows without ever really being heard:

> To the political, moral, and historical errors committed by the social or human sciences regarding man, whether individual or collective, when they ignore the exact sciences, and by both together when they forget the humanities, a strange blindness is added about what has been happening around us since the beginning of this same half century.
>
> While we were obediently repeating a dogma taken from sciences that were soft enough for us to doubt their relevance, the harder sciences, our law, communications … were constructing a new universe. (xi)

Serres reflects on a century of 'strange blindness', stating that historically (1900–50) we were preoccupied by universalism while living separate lives, then (1950–2000) we preached difference while paradoxically being brought closer together than ever by technology. He describes a time when the spectre of either mutual global destruction or the victory of one or other of the two competing ideologies of capitalism or communism, would prevail – either way, it was a singular outcome as the final victory of the unit over the multiple. The encroachment of a conclusive closing act is almost impossible to bear – the ultimate act of compression, the imposition of a universal model, however, had generated such a volume of residue and waste that it began to collapse under the weight of its own intensity. There was a purging of the sanctity of the unit, suddenly bereft of all the eternal objects required to sustain it. As a reaction, attention turned to more localized affairs, and the idea of the universal in the latter half of the twentieth century was displaced by difference. The universal had been rooted in dogma and domination, in a compressed view of truth readied for imposition. Multiplicity on the other hand began the process of expansion – beyond 'truth' and towards perpetual uncertainty.

Serres urges us to recognize this to and fro and to begin to think differently about difference, because while our gaze was averted the universal has returned – in the form of difference *as* universal. For Serres the second half of the twentieth century was characterized by myopia; a distraction or an illusion took place in front of our own eyes. While we were busy *looking* at the kaleidoscopic difference, interested parties were putting a new technological order together. It is this order that must now be addressed, and if the eye has been blinded then perhaps a new sensory mode of engagement is required. If, as has been argued so far, listening can generate fuller and less focused epistemological frameworks then we are compelled to consider whether the dominance of either the unit or the multiple were in fact historically accurate. What if neither of these propositions were strictly true, but simply symbolic/linguistic/compressed representations of a fluid state of affairs that in reality was far less defined? What if the unit and the multiple are aspects of a singularity that cannot be separated? If this is the case then the eye will always be afflicted with a degree of blindness, its gaze always limited and selective. The ear, however, can begin to accommodate the universe in both its aggregated and disaggregated modes, as both order and chaos.

In the midst of all of this, however, in the fullness of the universal soundscape, it is important to remember truths and certainties persist. As an example, Serres refers us back to geometry. Geometry defies the fate of unity that other modes of representation cannot match, and consideration of this signals the possibility of installing modes of engagement (a geometry of listening) that can encompass the multiple in a meaningful way without resorting to the singular logic of a unity. Geometry offers a set of 'universal' devices that are neither colonial in their ambition nor geocentric in their origin.

Similarly, the idea of compression can help us in this regard. It is different from the modern/postmodern positions insofar as it is embedded less in the linguistic turn and more in the sonic turn – residing not in the realm of constant uncertainty but in fluid pattern formation. Just because something is in a constant state of change does not mean that it cannot be known or understood (also asserting that things at the margins that are not always *seen* play a very significant part).

Serres calls for a new universal – one without dominance, and he sees in geometry and the language of mathematics a model for this. Geometry emerges at a time and in a place that can be located as Greece, but it does not emerge *from* Greece in any cultural sense. It is not 'Greek'. Its meaning does not relate to or refer to any locale. As a language it is not spoken or written, but rather is that which is true in all situations regardless of local variants or exemplars – it is in some ways a compressed linguistic system (approaching Wittgenstein's pursuit of an ideal language).

Serres's text makes the following points that are relevant here and require more detailed consideration.

The third time and the new map

'With the formation of the universe, a third moment appears to thought, one where the difference drawn on the globe or the map, those harlequin cloaks, mix by combining the ancient diverse paving blocks so as to strive toward the white of Pierrots costume … a plane or volume of geometry?' (xi).

This describes the designation of all the colours (as eternal objects) to create a beguiling but bewildering patchwork. In striving to bring the pattern to order, the colours mix/compress (as a form of lossless compression) until inevitably they return, each infused in the other, to white, each now indistinct from the other, and with that finite globality is achieved. Yet it remains true that white does not erase all other colours but contains them, and with this we are reminded that geometry can still restore and demarcate in full high-definition splendour, should we trust in it. For as Whitehead and Stengers have demonstrated, colour is a qualitative aspect of geometry in terms of distinction, as sound is in terms of movement and temporality.[1]

Just as colours combine to generate whiteness of Pierrot's costume, so sounds combine to generate, or return to, noise.

1 See Stengers 142–64.

The first construction of a universal

In praise of universality, as a principle not to lose sight of, Serres sets out a description of mathematics that traverses the globe, influencing most but not all, never imposing itself through military might, nor seeking to extend its cultural reach and influence. It is an example of universalism that serves all without having to conform to the orthodoxy of difference. If this describes geometry it also describes noise – a universal ground that pervades and supports difference in its fleeting transitoriness without recourse to atrophy as the victory of permanence.

Measure and reason

Talking of conflict driven by difference and its viral tendency towards generalized violence and war, Serres notes that rather than learning from persistent errors, humans continue to measure and redraw the world, according to some kind of, usually misplaced, reason, and usually on the basis of preserving difference and demarcation.

> What idealist arrogance in truth it is to think that we invent everything, according to the colour of our skin, the twisting of our tongues, and the gesticulations of our institutions! No, we're constrained to obey something other than ourselves, to obey an obligation that our measures don't dictate, inform, or show, to obey a demonstrated metric, a new universe, completely different from all our differences. What a blow to collective and cultural narcissism. (xiii)

This serves to bring into question the power of reason when situated within a nexus of mathematical universals, real factors as actual entities that do not depend on humans for their existence. The reasoning, measuring subject is reconstituted as an entity among other entities entangled in the mode of causal efficacy, and where even their passage into the mode of presentational immediacy cannot guard against error. Quite the contrary, error appears as a recursive characteristic. Yet it remains one that can be corrected (if we listen).

A key facet of Whiteheadian thinking, as it relates to the points raised by Serres, is that the reformed subject, though capable of creative interventions, is constituted in such a way as to embed them inextricably within the universe that sustains them and is bound by forces so fundamental that even the most creative subject is compelled to obey. Serres often characterizes this force as noise. Such a force compels without distinction or prejudice. Despite geometry having emerged from a specific point in time, like noise, it does not display any cultural or geographical characteristics. 'A disquieting strangeness: geometry then would go back to an origin, source or debut, to beginnings, without being attached to any root or flowering from any stem' (xiii).

As a universal, geometry is in a sense non-representational – it 'neither engraves nor traces any mark on any medium'. As such it does not seek to compress in the mode of presentational immediacy, or to impose, but only to express. It is therefore super-representational to deploy a phrase used earlier by Christoph Cox to describe music as a kind of 'pure mathematics'. As a language that makes no marks and speaks without prejudice, geometry may or may not describe the digital. If the digital retains the capacity to operate without leaving a trace, then it does so in terms of potential rather than actuality. Too often it serves to emulate its linguistic predecessors insofar as it seeks to isolate the unit from the multiple and impose a syntactical order on the universe.

If the universal has been lost to the specificity (localization) of both the hard and soft sciences, in geometry it remains. Through the inscription of the sciences and the arts (in the wrong hands) and their tendency to compress, much is cast aside, designated as residual and lost. But it can and must be recuperated as a source/resource of reconstituted knowledge and experience. Once again, the MP3 as a process of symbolic coding that allows constituent parts of a whole to be isolated and separated or redistributed or discarded stands as a reminder of what is possible, in the negative and positive sense. The ideal of the digital universal, as the simultaneous ability to compress and the capacity to expand, constitutes potential, but that potential has not to date been realized.

New habitat, our archaism

Serres describes his labours in geometry as a decades' long quest:

> to discover, to the eyes of the body and the mind, this new earth, a white or transparent box that was constructed starting from the field, the temple and the camp, slowly, from the spaces of violence, of prayer, of lustering or cleaning, of work, of defense or habitat to the page and the plane, the surface and volume, to all the varieties, the sum-earth of the acts of men whose complete purity replaced, from the Greeks on, the scalene plot of land and the oval planet, the vernacular local or nomad global, whose idea we then lost.

> Every single one of us has since inhabited the most immemorial of our universes, built from the spaces and times of Geometry. (xiv)

What is lost here? The promise of the universal geometry or that which it replaces? Both. The universal transforms into a 'jungle of sciences' and as a site of contestation and the embodied lived experience is subordinated to standardized measure – compression and expansion once more. It is possible however that there are good universals and that across the proliferation of particulars it is our job to find them within the immanent realm replete with modal features.

History doesn't always acknowledge the universal at its core and claims to leave what is proclaimed as belonging in the past behind, discarded, as it starts anew by virtue of a discontinuous turn. This historical process is central to Serres's thinking, as it was for Michel Foucault, to whom it is much indebted. It also resonates profoundly with Whitehead's history of symbolism as an emergent system where indeterminacy is resolved in the realm of error, or potential error as an opening for creative subjective intervention as the process moves from causal efficacy to symbolic representation.

Indeterminacy and determinacy, expansion, and compression are not zero-sum choices but interoperative elements of a single process. It is in their separation that problems arise.

> This indetermination has its limit in error, which the scientist forgets, but which the historian must reconstruct as the truth of an era; the historian is interested in dross, the scientist on the contrary is interested in the intuitions of genius without any impact on their epoch. Historical truth can change to waste, this latter on the contrary can be reactivated as truth. (xvi)

Indetermination's limit in error is evident when uncertainty is tamed, when an error is corrected. Until something has been 'proven' to be an error, indeterminism, possibility, and potentiality remain. Indeterminacy and determinacy are read by Serres through the figures of history and science. Science projects an irreversible trajectory based on its assertions and rejections of previous paradigms, often without a sense of contradiction or the indeterminate. Serres prefers to view history in all its guises as a chaotic system. He views 'all the preceding indeterminations as linked by a systematic space that recreates all possible chronic varieties' (xvii).

Such a spatio-temporal realm is the site of all potential events as understood from the perspective of the mode of causal efficacy. It is a site of temporality where the past may be recuperated and re-enter the system – waste reconfigured as resource. It is a noisy space wherein music will always succumb to its own temporality.

'For today we know how to decipher cuneiform tablets whose sexagesimal calculations resemble in a surprising manner the algorithmic procedures used by our computers' (xviii) – the return of thought from Babylon, scorned by the Greeks.

Such a temporality requires that we take account of the finely fibred, foliated, weft of time. The *science of history*, and therefore of time – how change occurs, moves, patterns, continuously and smooth or discrete and abrupt, or an entanglement of all the features, themselves operational in time and space. Such a celebration of complexity and the need to include everything can lead to problems, however, and ultimately to the knowledge and understanding of nothing – to submission in the face of bewilderment. Geometry, or more specifically a geometry of listening, serves to counter this.

Invention and tradition

'Thus, all won ground illuminates and occults the history of science, in aleatory rhythms: the current invention discovers precursors, while forgetting former origins destined to become dross. Often in crisis, mathematics is always in the process of resolving it' (xxi).

The history of mathematics is one of advance and return of rejection and recuperation, always reminding us of the complex nature of history itself, but equally reassuring us of the universal nature of certain principles. Whereas we lose touch with our past and the knowledge it contains across all aspects of our existence, we can be assured that in mathematical terms our knowledge is sound: 'we know without any risk of being wrong what they [the Greeks] were thinking when they devoted themselves to apagogic proof. No equivalent in knowledge or historical information is known' (xxii).

Stages or crises

'The progressive extension of the mathematical field, the continuing purification of its concepts, the always strengthening power of its methods, the forward movement toward a mathematicity conceived as horizon give us to think an evolving form that's connected but punctuated with stages, steps, or crises, global reorganisations of a knowledge transmitted without any loss, therefore incessantly accumulated' (xxiii).

Like the idea of repetition – repeating to the nth degree – mathematics repeats in a particular way – not a single point of origin but a complex historical trajectory, with breaks, returns, recuperations, and linguistic expression, but always mindful of the solidity of the ground on which it is born, and reborn. Yet it is a particular kind of accumulation that it displays, one of intensity rather than scale (compression and rarefaction). The history of mathematics like compression is a filtering process.

Percolator

> What a wonderful and luminous technique of archaeological research then this filtering for purity is, a filtering achieved by the very movement of mathematics' progress! Every point along its course discovers evidence of the origin carried up to there and abandoned through the contemporary filtering – yes, well known evidence of pre-history: as in astronomy, you can receive information from worlds that no longer exist. (xxvi)

That there may be much to learn from that which has been discarded is a central tenet of the argument being constructed here.

On the excess of information

> We periodically hear uttered the fear that the accumulation ok knowledge leads as inevitably to barbary as its very absence; science ought to collapse beneath its own productive proliferation. This amounts to believing that the advance in what we know recuperates the totality of previous knowledge: through an accumulative process, the encyclopaedia would snowball on itself.
>
> For mathematics at least, things don't happen like this because it filters its heritage rather than taking it up in its entirety; better, it takes it up by filtering it. *(xxvii)*

element of truth in a transient and dynamic universe. Historically, the conjoining of sound and geometry has invoked a degree of mysticism, but it is a mysticism that is formed of precise mathematical patterns. The idea of a sacred geometry builds upon a series of mathematical models including Sumerian base 60, the Fibonacci sequence, the resolving to harmony of Pythagoras, and Mandelbrot's fractals, that ultimately form what we understand as music.

The intention here is not to repeat, or engage in any detail with these mathematical models, but to note their significance in terms of combining natural processes and rational interpretation to form logical conclusions, as Serres does with his assessment of geometry. What is important is not evidence of the sacred, but the presence and dynamism of patterns and rules.[3] It is this simultaneous interaction between transitoriness as dynamic and permanence as pattern and rule that both explains and calls for a geometry of listening.

The specific structure of this musical geometry has been set out in detail by Dmitri Tymoczko.[4] He, along with Clifton Callender and Ian Quinn, devised a new way of analysing and categorizing music by deploying the deep, complex mathematics that is identified as being enmeshed in its very fabric.

> Understanding music, the authors write, is a process of discarding information. For instance, suppose a musician plays middle 'C' on a piano, followed by the note 'E' above that and the note 'G' above that. Musicians have many different terms to describe this sequence of events, such as 'an ascending C major arpeggio', 'a C major chord', or 'a major chord'. The authors provide a unified mathematical framework for relating these different descriptions of the same musical event. They describe five different ways of categorizing collections of notes that are similar, but not identical. They refer to these musical resemblances as the 'OPTIC symmetries', with each letter of the word 'OPTIC' representing a different way of ignoring musical information – for instance, what octave the notes are in, their order, or how many times each note is repeated. The authors show that five symmetries can be combined with each other to produce a cornucopia of different musical concepts, some of which are familiar and some

[3] The distinction between music and noise is mathematical form. Music is ordered sound. Noise is disordered sound. Music and noise are both mixtures of sound waves of different frequencies. The component frequencies of *music* are discrete (separable) and rational (their ratios form simple fractions) with a discernible dominant frequency. The component frequencies of *noise* are continuous (every frequency will be present over some range) and random (described by a probability distribution) with no discernible dominant frequency. https://physics.info/music/.

[4] See *A Geometry of Music: Harmony and Counterpoint in the Extended Common Practice* (2011).

of which are novel. In this way, the musicians are able to reduce musical works to their mathematical essence.

This discarding of information is clearly of great importance here. It demonstrates that in music there exists both temporal unfolding in real time and extremely precise geometrical patterns whose description can be compressed to produce clear analytical expressions or rules. Spatializing audio in this way involves assigning musical points in topological three-dimensional space where those points are in motion. Whitehead understands this as 'point free' geometry[5,6] – not describing an absence of points but a set of mobile, and dynamic, rather than static, relations. It might better be understood as a non-geometry insofar as it describes a process through which perception in the mode of causal efficacy subsequently resolves towards geometry in the mode of presentational immediacy and onwards through its progression towards the symbolic, where mathematics and culture entwine. What is important is that harmonic relations are grouped into topological regions rather than fixed points. If such a description is accurate, we return once more to the Whiteheadian universe.

Concerned predominantly with regions rather than points, Whitehead initially deals with what is known as mereology (the relation between parts and the whole), and later expands his scope to a form of mereotopology (the invariant relationship between parts within a whole whose shape and size can change continually). Parts, wholes, units, multiples, and locations are all factors that are relative to one another within dynamic systems that include social, political, economic, technical, and aesthetic formations. In all modes, systems are subject to phases of compression and expansion as they resolve to order while simultaneously being compelled to resist formal classification.

The mereological and the mereotopological are present in music, and particularly in digital formatting of sound where data points within a musical composition can be stretched and reorganized into different, but related, versions. Maguire's *Ghost in the MP3* demonstrated this. In his case parts of a whole are isolated and removed only to be reassembled, where a degree of entropy is present, outside of the original configuration. The completeness of the

5 https://www.scien ceda ily.com/relea ses/2008/04/08041 7142 454.htm.

6 See G. Gerla and M. Annamaria (2008), `Mathematical Features of Whitehead's Point-Free Geometry', *Handbook of Whiteheadian Thought. De Gruyter*. https://www.degruyter.com/document/doi/10.1515/9783110333299.2.119/html?lang=en

original song, in a way that draws us back to Whitehead, functions as a system that is open – to both incoming and outgoing data. No sooner is music (as organized sound) confined to notation and understood in relation to compressed geometrical (or digital) codecs or formula however, then it does something unexpected. Through changes in context, proximity to other forms, technical error, playback conditions, audition, or challenged by new designative forms, or by noise, it begins to defy its own forms or representation. It compresses and expands so fluidly as to make it uniquely able to reveal the processual nature of all relations, entities, and events.

Similarly, with point free geometry Whitehead attempted to come to terms with the interconnection of all things, describing a process of ordering the universe into sensible units, either units of measurement, units of spatial division, or units of perception within a 'bewildering' whole. Central to this is the idea of patterns and repetition, but also the unsettling of the notion of location as a fixed point. This does not constitute an abandonment of geometry but is rather an attempt to accommodate movement and process into its precise measuring and locating coordinates. As such it resides comfortably within a sonic economy and supports the idea of a geometry of listening.

Whether substantivalist or relationist, from noise, sound and music emerge as events that, fleetingly at least, possess substantial geometric qualities that can be said to be real. So much so that they reveal the process in states of energetic transformation and convolution. They also provide the ground for rules, but also for error and with-it creativity. But it is not the creative potential for error that should guide us. We should not seek through indeterminism the opportunity to compress and generate waste, but to correct and include where apposite – through a rebalancing of the ratio between mathematics and culture, the universal and the particular, as they move through the 'bouquet of time'.

The issue of relations and how they work in reality is important and relevant here. Much of what has been discussed thus far has been oriented around a challenge to the classic subject/object dualism and the establishment of a monistic ground, or a one substance cosmology. However, here too differences of approach and emphasis emerge. The distinct positions of F. H. Bradley, Whitehead, and William James, are worth noting. They variously describe either a non-relational absolute (Bradley) or, as was the case with James and Whitehead, 'novel forms of metaphysical pluralism in which a concrete relatedness becomes the essential defining characteristic of each actuality in a creative universe' (149).

This is a complex matter: multiples and singular units, multiplicity as a singular and singularity as multiplicity. Monism as a singular set of relations or the more totalizing assertion that the universe is truly a singular entity within which discrete entities are either mere appearances or facets of a greater whole, understood in part through a mode of metaphysics that is ultimately unable to reflect on itself. Once again, we find ourselves presented with a debate that cannot be adequately settled here, but which frames the argument and acts as further justification to move out of the realm of appearance, as predominantly ocular, and into the sonic realm of noise, sound, and music. 'The main point seen by each is a certain injustice done to concrete experience once analysis has cut into what is essentially alive and harmonious' (154).[7]

The idea of analysis 'cutting into' is important in relation to the mode of compression where the fullness of reality and experience of it, whether as a set of relations or a more pure form of monistic absolute, is symbolically and linguistically segmented in order that some sense of relief from chaos and bewilderment, such that it includes the self as raging torrent of noise, can be achieved. This process of segmentation is largely symbolic and linguistic. James argues that the intellect can only handle a type of retrospective 'patchwork' or 'post-mortem dissection' of reality because perceptive capacities cannot accommodate the totality of 'the cutting-edge of immediacy', and Whitehead identifies in the static nature of the subject/predicate proposition an inability to account for dynamic flow. Both describe modes of compression where the universe is variously divided up into comprehensible parts, and where any changes to those parts or relationships between parts is halted. In doing so that which is 'essentially alive and harmonious' is terminated and rearticulated according to a set of geometrical points, symbols, and signifiers. Yet this murderous and abrupt intervention is part of the system, and simply to pit harmony and symbolism against one another is counterproductive. Instead, it is necessary to rethink the nature and the extent of harmony in relation to the different kinds of geometry, or indeed the difference between a geometrical and a dynamical understanding of the universe and of spacetime.[8] This must be done in relation to physics and metaphysics, things that change and things that stay the

7 Leemon McHenry, 'Bradley, James and Whitehead on Relations', *The Journal of Speculative Philosophy*, New Series, vol. 3, no. 3 (1989), pp. 149–69.

8 See James Read (2020), *Geometrical Constructivism and Modal Relationalism: Further Aspects of the Dynamical/Geometrical Debate.*

same, points and measurement, or a more dynamic sense of place, division into specific fixed units with borders and thresholds, or more amorphous topological regions. These differences can be accommodated and modulated sonically to produce new arrangements and effects. Sound can both conform to geometry and can also behave in a way that cannot be contained and easily bounded.

If noise, sound, and music are more appropriate modes of engagement with the non-relational absolute of Bradley, as they are in terms of revealing the process of causal efficacy, then the visual dominates in the mode of presentational immediacy where form and its qualities arise from differentiation and relative distinction. In combination as modes of experiencing, as Whitehead goes some way towards establishing and as does Serres (in *The Five Senses*), they begin to account for a universe in which humans, as well as being distinct and unique, are also integral elements of what might be described as a super-relational ground. There is not a simple choice to be made between the visual and the audible. Rather, aspects of each provide, as modes of experiencing, different but always related levels of understanding.

Emergent and dynamic processes, the connection within and between things, their capture, measurement, as representations, either fixed or in motion, in spatio-temporal terms are under consideration here.[9] Dualism, monism, pluralism, all seek to understand and establish the ground or grounds on which reality subsists. That the real exists and does not depend on the presence of a rational subject to substantiate it is accepted here. However, whether it is a real that exists ultimately as a singular monistic absolute as Bradley asserted, or as space inhabited by a plurality of ultimately withdrawn and unknowable (relational at one level but entirely independent at another), as Harman has proposed, and indeed whether Bradley and Harman are as far apart from one another as it would at first seem, are all important factors, but ones that regrettably cannot be resolved here (but are certainly a matter for further investigation). What can be established is the significant contributions made by both Whitehead and James

[9] But how are we supposed to measure how wise or prudent we are in choosing the instruments of measurement and interpreting the findings? Modern literature is full of references to the dehumanizing side of measurement, as exemplified by the character Thomas Gradgrind in Dickens's *Hard Times*, a dry rational character who is "ready to weigh and measure any parcel of human nature, and tell you exactly what it comes to," yet loses track of his own life.

How can we keep an eye on the difference between ontic and ontological measurement, and prevent one from interfering with the other?

to the systematic modelling of relations and the ways in which they constitute our experience of the world, and therefore our knowledge of it.

> Standing between the radical pluralism of Hume and Russell and the radical monism of Spinoza and Bradley, James and Whitehead have both defended what James called the legitimacy of the notion of some, for as he argued: '… each part of the world is in some ways connected, in some other ways not connected with its other parts, and the ways can be discriminated'. (McHenry, 1989: 155)

In order for things to be in-relation they must first be distinct and independent. It is this distinction that is at stake in the differences between Bradley, James, and Whitehead. From an ocularcentric perspective the problems present themselves in particular ways: appearance and reality, the difference between primary and secondary qualities, and so forth. In sound however things may be different. The figure of noise can accommodate all of their stated positions, especially if we add to the legitimacy of the 'some' that of the 'sometimes'. Noise is then both a singular substance and a set of relations – both wave and particle.

By way of a slight diversion this dual aspect singularity is also what is described by Michel Foucault, who importantly confronts the difficulty in locating dynamic events from a specifically political perspective. Foucault informed us of the difficulty inherent in attempts to locate aspects of our experience that are dynamic and occur perhaps within regional (point-free) rather than specifically point-based locations. As such they are understood here to emerge as sonic configurations.

In *The Archaeology of Knowledge*, Foucault gives an account of what he refers to as a statement. The statement is a rare but very significant phenomenon that takes its place among the 'noisy' world of propositions and phrases. Its rarity signifies an almost improbable occurrence, an actual entity as event that inheres from an almost limitless set of potentials. A statement does not always reside within familiar patterned structures or spaces. Statements can form out of processes that lie outside of formal modes of structuring. They emerge from a confluence of often disparate factors. So much so that they are not always easy on the eye *or on the ear* and therefore cannot be constrained by the giving of examples (giving examples is a form of compression, or a crude algorithm). They are the loci of power as both discourse and practice – they matter, they make a material difference, they have impact, and in a compressed linguistic mode they are made from 'noise'. What is important about statements is that they attain a degree of influence as they are instantiated through language as a normalizing framework. They are subject to a strange and paradoxical geometry, sharing aspects in common with both Whitehead and Serres, where their location is indistinct, but its impact is quite precise, and as such demonstrates characteristics that are also present in acoustic and sound art practices.[10] As a linguistic event that is a distinct signal in an aspatial sea of noise the statement, Deleuze suggests, emerges through a process of 'atonal logic'. By this is meant a mode of revealing that does not declare itself to 'conventional' approaches that draw on fixed positionality or the consistent exchange between signifier and signified.

The idea of atonality is important here. A statement is a thing of rarity. Amid all the fervent activity and unlike propositions and phrases that are ordered vertically and horizontally, statements cannot be located on a grid of predictability (with its regular patterns). Instead, they are found to reside in the less ordered and often unpredictable realm of the 'diagonal', and even when they are tracked down their superpositional qualities enable them to quickly avoid our gaze. Statements are diffuse, so much so that diffusion, as a practice and a model/technique for initiating new forms of engagement/socio-economic sensibility, as cultural diffusion, political diffusion, is essential if they are to be adequately identified and understood. Is it time to think these ideas anew, to echo and amplify Foucault's call, this time in a specifically sonic mode?

The distinction or the contrast between a precise point-based geometry and a point-free approach that identifies regions and morphing topologies has been a feature throughout the discussions so far. Fixed coordinates or dynamic relations, the visual and the audible, measuring and not measuring, tonality and atonality. As such, and having already drawn extensively on Whitehead and Serres, Deleuze and Guattari are also significant in this regard, specifically their concepts of the *Body without organs* and the *smooth and striated* (musical model). Like Foucault's immeasurable contribution, these ideas, concepts, refrains, have received an enormous amount of attention and are the subject

[10] Acousmatic practices specifically, insofar as they detach the sound from its specific locatable source, using a diverse array of techniques, are able to critically engage with notion of relationism and geometry. By isolating and rearticulating certain specific sounds, the absolute totality of Bradley, by cutting into what is 'alive and harmonious', is disrupted. At their best such practices can reveal the relational process rather than subjecting them to static representation as is the case with other mediating practices.

of a great many academic texts, conversations, and events. The intention here is not to add significantly to that academic corpus, but rather to place them, as samples, within the larger arrangement of ideas under consideration here.

The concept of *the body without organs* describes a scale of extremes that excludes the received wisdom of a universal patterned normative state and the negative notion of the lack that places humans into a gridded mode of standardized measurement and expectation that is grounded in modern psychoanalysis. In its place a new mode of investigation and existence is proposed based on the interconnection of organic and inorganic beings that both attract and repel, making positive contributions to an ever-evolving ecosystem – a sonic economy where order and chaos are modal and uncertain.

In *Anti-Oedipus* (2004a) Deleuze and Guattari write:

> Lenz has projected himself back to a time before the man-nature dichotomy, before all the co-ordinates based on this fundamental dichotomy have been laid down. He does not live nature as nature, but as a process of production. There is no such thing as either man or nature now, only a process that produces the one within the other and couples the machines together. Producing-machines, desiring-machines everywhere, schizophrenic machines, all of species life: the self and the non-self, outside and inside, no longer have any meaning whatsoever. (2)

The schizophrenic on a walk engages with a different kind of mapping, following an atonal map where the desiring machine generates connections and acquires territorial familiarity through connection and experimentation where the reward mechanism is triggered by predictive uncertainty rather than guarantee. This constitutes a cartographic mode as *method* of engagement. In its temporal unfolding it is a sonic method, where the very notion of method is prefaced on the presence of movement rather than stasis. By way of a helpful definition, Rancière defines a method as a form of travel that 'continuously discovers new landscapes, paths or obstacles which oblige to reframe the conceptual net used to think where we are' (Rancière 2003: 120). Accordingly,

> what he does himself is to construct a moving map of a moving landscape, a map that is ceaselessly modified by the movement itself' (ibid.). To ascertain the characteristics of this landscape requires more than the tools of the prospector trying to extract minerals from the soil. To grasp this 'moving landscape' requires developing a sense of orientation to figure out where you are, where you are going, and where you can go with it, all of which cannot be represented

as a still image. To 'map' it is neither to produce a flat cartography, nor to freeze it in time, but to get it on with its movement, in order to recreate a method's 'topography of the thinkable', which 'is always the topography of a theater of operations'. (Rancière 2003: 19)

To always expect the unexpected in this 'theatre of operations' is not to negate the idea of patterning completely, but to embrace a different kind of pattern – one rich in difference and potential.[11]

To this end we move to the *smooth and striated* – as spatial qualities, modes, and relations. The ways in which they resonate and cohere continues the trajectory from Bradley through James and Whitehead where both absolute and relative qualities are evident as a kind of wave/particle duality. The idea of a modal interaction between states of order and chaos, as compressed and uncompressed realities and perception of realities that are themselves interrelated, is a recurring feature of many of the arguments examined thus far. The smooth and the striated spaces, as described by Deleuze and Guattari, are presented as either nomadic or sedentary, dynamic, and free to roam along the atonal, diagonal, or static and fixed according to point-based coordinates. Yet the two spaces are not contradictory. They exist in relation to each other in a more complex way than can be accounted for by simple opposition. The two spaces exist only in mixture. 'But the de facto mixes do not preclude a de jure, or abstract, distinction between the two spaces.' The two spaces do not communicate with each other in the same way. It is not a straightforward process of communication with pre-agreed protocols. Organizing or mapping noise and chaos, as the smooth, on a grid using the grammar of striation is not an option – the map as a mode of compression is not best placed to locate that which is uncompressed.[12]

[11] For substantivalists, space and time exist as ontologically autonomous entities in the world; for relationists, by contrast, space and time are (in some way to be articulated) ontologically reduced to (relations between) material bodies/fields. This much is extremely well known in the philosophical literature; what is slightly less known is one particular historical sub- distinction *within* relationalism – that is, the distinction between 'modal' versus 'non-modal' relationalism. According to modal relationalism, space and time are ontologically reduced to (relations between) matter bodies/fields *in all possible worlds*. By contrast, according to non-modal relationalism, space and time are ontologically reduced to (relations between) matter bodies/fields *in the actual world*. (Read, 2020:14).

[12] This is because we are talking about two different kinds of compression, lossless (a compact intensity/event that contains the full statistical curve), and lossy (where in order to achieve optimum reduction waste is created). The smooth or the body without organs can be compressed, theoretically at least, into an ideal state of perfection, but in reducing it and producing waste, striation reigns.

> This raises a number of simultaneous questions: the simple oppositions between the two spaces; the complex differences; the de facto mixes, and the passages from one to another; the principles of the mixture, which are not at all symmetrical, sometimes causing a passage from the smooth to the striated, sometimes from the striated to the smooth, according to entirely different movements. We must therefore envision a certain number of models, which would be like various aspects of the two spaces and the relations between them. (Deleuze and Guattari, 2004b: 524)

The relation between them refers us back to compression and expansion where compressed and ordered units of information are distinct but not separable from an expansive multiplicity of potential outcomes. It also refers us back to the visual and audible fields.

> Pierre Boulez was the first to develop a set of simple oppositions and complex differences, as well as reciprocal non-symmetrical correlations, between smooth and striated space. He created these concepts and words in the field of music, defining them on several levels pre- cisely in order to account for the abstract distinction at the same time as the concrete mixes. (2004b: 527)

This could be describing the interplay between causal efficacy and presentational immediacy, as respectively smooth and striated spaces.[13]

> The smooth is a nomos, whereas the striated always has a logos, the octave, for example. Boulez is concerned with the communication between the two kinds

13 In the section on the musical model, Deleuze and Guattari discuss musical composition, and they do so by referring to Boulez (from whom they take the terms smooth and striated), a figure perhaps best known for his extension of serial techniques into musical domains other than pitch (e.g. timbre, duration, intensity) and for the freedom he would sometimes give performers to determine their own route through the score. Here, in a short amount of space, Deleuze and Guattari again lay out several differences between the smooth and the striated. There is, for example, the distinction between nomos and logos. 'The smooth is a nomos', they write, 'whereas the striated always has a logos, the octave, for example'. The difference here is between a kind of reason-based, binding law (logos) and a variable practice of custom (nomos). Perhaps, however, custom is too restrictive a term for the difference they are trying to articulate, for it, too, can be extremely rigid. Indeed, while it might be tempting, at first glance, to view logos as being a restrictive and absolutely negative force and nomos as being a liberating and positive force, Deleuze and Guattari have no such delusions. In discussing the differences between societies based on logos and those based on nomos (in the section on the physical model), they write, 'What they [i.e. so-called primitive societies] have instead [of logos] is a law of the nomos regulating a continuous variation of activity with a rigor and cruelty all its own (get rid of whatever cannot be transported, the old, children).' Bearing this in mind, the core difference here is between a rigid enforcement of laws that dictate and pre-determine actions and a variable practice that opens up the possibility for acting differently in un(-pre-)determined ways (Faul, 2018).

> of space, their alternations and superpositions: how 'a strongly directed smooth space tends to meld with a striated space', how 'a striated space in which the statistical distribution of the pitches used is *in fact* equal tends to meld with a smooth space'; how the octave can be replaced by 'non-octave-forming scales' that reproduce themselves through a principle of spiralling; how 'texture' can be crafted in such a way as to lose fixed and homogeneous values, becoming a support for slips in tempo, displacements of intervals, and *son art* transformations comparable to the transformations of *op art*. (2004b: 527)

As a means of putting this in context and amplifying the significance in terms of sound it is worth returning to the idea of the sonic image and the 'accent' of the word as a means of mapping specific spaces:

> An accent should not be understood as a pronouncement of place of origin, but rather as equivalent to layers of metadata. Accent is an indication of the voice as network, a network with nodes in various places, times and identities. An accent is formed through and in communication with others; it is relational, in the sense that we hear not only the speaking subject but traces of their past and present interlocutors. The voice has a tendency to mimic the accents of those it speaks and relates to so that the sound of an accent is possessed to some degree by every person it has come into contact with. When the accent is heard as a network and not a birth certificate, we hear it as a biography of migration, as an irregular and itinerant concoction of contagiously accumulated voices, rather than as an immediately distinguishable sound that avows its defined roots within the confines of a nation-state. Such irregularities should provide proof of the applicant's need for asylum and stability. But instead, the sounds of vocal migration, which have the sonic quality of geographic irregularity, are cited as grounds for refusal of asylum applications, as LADO promotes a system of vocal governance that deems speakers with mixed and multiple accents as dishonest. Yet in the governance of accents, the border is a visual demarcation that dissects a sonic plane. And in these images, I am trying specifically to invert that, to show that sounds cannot be so easily dissected by the colonisers' cartographic rulers and pens. It is through this sonic image that I propose strategies for mapping the voice that are more faithful to the way the medium bleeds through histories, bodies and borders. The attempt here is not only to render an image of a sound but rather to create an image that behaves like sound itself.[14]

[14] https://infrasonica.org/en/audiblematter/thesonicimage.

Points, regions, the absolute and the relation, the smooth and the striated, and the ability of sound to bleed through borders bring us more explicitly to the concept and to the practice of mapping, or more importantly to cartography as a mode of (sonic) compression.

Cartography is a science of scale, and reduction, and sometimes omission – once more the removal of 'unnecessary' data without compromising the intended message or meaning. Neil Gaiman (2007) says:

> One describes a tale best by telling the tale. You see? The way one describes a story, to oneself or to the world, is by telling the story. It is a balancing act and it is a dream. The more accurate the map, the more it resembles the territory. The most accurate map possible would be the territory, and thus would be perfectly accurate and perfectly useless.

This idea was also explored by Lewis Carroll in his novel *Sylvie and Bruno* and was further expounded upon by Borges in 'On Exactitude in Science',[15] as referenced by Baudrillard in his *Simulation and Simulacra*, and again in *Don Quixote*.[16] The story in this instance describes an attempt by Mennard to recreate the text of Don Quixote without copying. Mennard sets out to encounter a universe of possibilities that statistically at least, supports the possibility of

[15] Jorge Luis Borges, 'On Exactitude in Science', in *Collected Fictions* (2000), translated by Andrew Hurley

> In that Empire, the Art of Cartography attained such Perfection that the map of a single Province occupied the entirety of a City, and the map of the Empire, the entirety of a Province. In time, those Unconscionable Maps no longer satisfied, and the Cartographers Guilds struck a Map of the Empire whose size was that of the Empire, and which coincided point for point with it. The following Generations, who were not so fond of the Study of Cartography as their Forebears had been, saw that that vast Map was Useless, and not without some Pitilessness was it, that they delivered it up to the Inclemencies of Sun and Winters. In the Deserts of the West, still today, there are Tattered Ruins of that Map, inhabited by Animals and Beggars; in all the Land there is no other Relic of the Disciplines of Geography. Suarez Miranda, *Viajes de varones prudentes*, Libro IV, Cap. XLV, Lerida, 1658

[16] It must be noted that Borges is here granting these concessions in order to refute idealism, but he is also fascinated by the possibility of discovering an ancillary value in 'practicing excess', in setting up within his fictional domains the opportunity to 'search for unrealities'. If the scientific method of the modern age is to be dominated by loss of method and uncertainty principles, then Borges finds no difficulty in developing fictional characters who pursue unprovable conjectures, especially those that relate to 'unreal' or quixotic ventures. Readers and critics who accept the scientific worldview of the twentieth century are far more likely than their seventeenth- and eighteenth-century counterparts to take such ventures seriously and less likely to dismiss them out of hand as comic buffoonery. https://www.borges.pitt.edu/sites/default/files/0709.pdf.

exact replication. If one could inhabit the milieu of another, would it not be possible to engender the same causal efficacy, like a scientific experiment that can be repeated? Possibly, were it not for the mode of presentational immediacy and the drive to compress. The raging torrent of information that is received is ordered in such a way as to move it towards the symbolic, and it is this symbolic representation that is shared between subjects, not the fullness of possibility. So much so that each symbolic concept, even if it is unique to each individual in the mode of presentational immediacy, retains enough of the mode of causal efficacy to facilitate a shared understanding in common – as Wittgenstein described, but at a level of resemblance that is not repeatable.

In another Borges text we encounter the Library of Babel as an algorithm:

> The universe (which others call the library) is composed of an indefinite, perhaps infinite number of hexagonal galleries … Idealists argue that the hexagonal rooms are the necessary shape of absolute space, or at least our *perceptions* of space. They argue that a triangular or a pentagonal chamber is inconceivable. (Mystics claim that their ecstasies reveal to them a circular chamber containing an enormous circular book with a continuous spine that goes completely around the walls. But their testimony is suspect, their words obscure. That cyclical book is God.) Let it suffice for the moment that I repeat the classic dictum: *The Library is a sphere whose exact center is any hexagon and whose circumference is unattainable.* (2000: 66)

The library contains books that comprise all the combinations of all the letters available, sometimes constituted as jumbled nonsense and sometimes as exhibiting fleeting coherent sentences. Possibly somewhere in the almost infinite reaches of the library lies a book that contains all the secrets insofar as it is entirely intelligible. Somewhere in the library truth must exist it was thought, and that truth/vindication was sought: 'The vindications do exist (I have seen two of them, which refer to persons in the future, persons perhaps not imaginary), but those who went in quest of them failed to recall that the chance of a man's finding his own vindication, or some perfidious version of his own, can be calculated to be zero' (2000: 70).

There is no ultimate truth, no book of secrets. Faced with this prospect humanity calls upon itself to, or is compelled to, map, reduce, and condense. But how can a universe of endless opportunities be mapped when a map is a reduced version of reality with detail missing? What is required to guarantee meaning and what can be dispensed with?

In the digital realm the entirety of the world is being mapped, twinned, replicated, and modelled. Images/representations/datasets make manifest some of the most complex aspects of existence – modes of experiencing, perception/conception, ontological status, dimensionality, and aesthetics are all encompassed within digital discursive regimes. These in turn have very significant political and economic repercussions in terms of who gets to speak and about what, and how new channels of communication can support enhanced participation – or not. The perceived lack of engagement and inability to influence rapid change is generating unprecedented levels of inequity, both economic and in terms of access to information. How can we alleviate this collective anxiety in the face of multiple global crises? In considering this it is important to rearticulate renderings of the real and objective as accessible assets that are part of an aesthetic regime that generates new possibilities for people to see (in a new light, producing a sonic image), to hear and to participate in the construction and reconstruction of narratives in all their orientations. The symbolic realm, as a mode of compression, must be sonified, revitalized by the addition of sound so as to make it accessible and meaningful in a contemporary context so that narratives can be critiqued and reassessed on an ongoing basis rather than forming permanent reference points. Faced with rapid social, cultural, and technological change, populations more than ever need to be reassured that external circumstances are not primed to overwhelm and exclude them, as by-products of compression. It is essential to enhance/expand engagement with spatio-temporal experience, to produce a better map of history (as temporality) and our place in it. Citizens as immanent and reformed subjects are integral to the movement of history and should not be misplaced victims of forces perceived to be beyond their control.

The production of digital models, as symbols, can and should serve to give prominence and visibility to some key spatialities in such a way as to highlight their transitory nature and allow users to interpret the changing nature of the real through an appropriately technological lens. The use of digital technology to produce content and to disseminate it widely should broaden awareness of the material universe. More importantly though it must engage subjects, participants, and users in such a way as to locate them and their interpretive skills and experiences as central to, rather than separate from, the process of change. Broadening awareness of the changing nature of material reality and objects, challenging their status as permanent monuments to the past by placing them firmly in the present, and orienting them towards the future requires a temporal dimension. The inclusion of sound as an important aspect of the real is essential in this regard insofar as it goes beyond the current ocularcentric state of the art which is prefaced upon objective historical certainty and impermeability. Sound on the other hand brings to the fore transitoriness, the fleeting nature of interaction and the celebration of change, possibility, and potential.

The systematic description of interacting levels of energy and information exchange that Serres sets out complement perfectly Whitehead's process philosophy, and just as thermodynamics gave rise to new conceptions of systems, so advances in digital technology have made us think once more. Related instruments allow us to see and listen in new ways, with new eyes and ears. Maybe a signal can now be discerned from a 'vague and fluctuating cloud, from a multiple halo humming and buzzing at random'.

Concluding remarks

The sum total of human thinking and its efforts to put ideas into practice has been compressed into divergent ideological units. Filtering out inconvenient or distasteful elements, they seek to model and map the universe in ways that serve to privilege a statistical average of the human population. However, like the search for the perfect book in the Library of Babel, their efforts have largely been in vain. The proliferation of global crises abounds, and many technological prophecies have proved unfounded. The intention of this book has been to articulate some of these ideas and position them in new configurations relative to one another, in a way recomposing and remixing elements to forge new modes of experiencing that are better equipped to deal with contemporary realities, and which are ultimately orientated towards truth.

Taking stock, striking a balance between the arts and humanities and the sciences – hard and soft, has required cooperation between disciplines to craft and test a workable mode of engagement that embraces noise and chaos rather than seeking to quieten and order in absolute terms. Combining critical theory with art, acoustic composition, and science, it is possible to enact a new model of political economy – or a post political economy (a sonic economy). Ideas if left unattended can easily pass us by, taking up residency in forgotten realms. Such ideas often form within a context of academic discourses that are not effectively communicated. As a largely theoretical text it is recognized that this book inhabits this contextual space, drawing as it does on the work of others, predominantly Whitehead and Serres, in this instance, to generate convergent and divergent moments/confluences/intensities that operate within rather than outside the existential terrain of both aesthetics and political economy. The overriding aim has not been to offer a new model or map, but to stress the point that models, maps, and language in all their forms, serve to compress, and often to exclude a significant quantum of data. This is not in itself being challenged, and as has been demonstrated, it forms a natural part of a system of interaction and information exchange. What is being challenged is the extent to which this compression, as it operates within the mode of presentational immediacy, generates waste that is consigned to peripheral realms. The symbolic, the representational, and the eternal deferral of meaning are constitutive of a 'self' abstracted from the real, disconnected from the mode of causal efficacy, and ultimately responsible for an erroneous view of the subject and its complicity in a universe of immanent relations. In this volume ideas do not enter forgotten realms to wither but rather take their place within a complex ecology of forces and pulsions. Such an ecology now includes technological and technologically enhanced modes or perception. Extra-dimensional and extra sensory data from technologies such as photogrammetry, lidar, hyperspectral imaging, Eulerian video magnification, in the visual realm for example, and digital signal processing in the audio realm, are initiating a shift in the subject/object paradigm. With this shift come aesthetic and political implications. The inclusion of 'peripheral' data into experience must extend to all areas of life. If ocularcentrism delivered a compressed and largely visual framework wherein the 'inconvenient' was simply dispensed with, newer forms of perceptive fields invite inclusion and complexity in both audio and visual realms.

To accommodate this and to counter the assumptions that proceed from an over-reliance on the received wisdom of compressed, symbolic representations and abstractions as units, there is a need to rethink the idea of the multiple, even when, or especially when, it occurs beyond our view. Central to this has been Michel Serres and his notion of 'the multiple' (in *Genesis*) as a descriptor of both the real and the virtual, and his advocation of listening (in *The Five Senses*) as a mode of engagement that is better able to account for the global (dis)order thrown up by the proliferation of 'the digital'. In recasting some of these (and other) ideas, questions are inevitably raised around political engagement and policy frameworks and design (drawing on Zygmunt Bauman's ideas in *Wasted Lives*) and the distribution of the sensible. To address these questions, it has been necessary work through a range of modes of engagement that not only accept but celebrate the diffuse nature of sensory experience ('the multiple') and in the process offer up new models and political processes that are appropriate and inclusive.

> The most fertile methods today concerning the mythical text in general are regulated by an algebra and, more precisely, by a combinative algebra. There exists to begin with – or, better yet, it is possible to constitute – a set of discrete elements, of units. Out of this reservoir circulate combinative sequences that can be mastered. Hence the theory of musical forms that is certainly the most general available organon, both practical and constructible, for these operations. This algebraic method is, to my mind, a local realization three centuries later of the Leibnizian dream of an alphabet of human thoughts for which its author had forged an ars combinatoria – first invention, precisely, of combinative algebra as well as of a logic of the note, of any discrete note. From which Leibniz derived the idea that music was indeed the language closest to the universal language, or to the mathesis universalis. This was an idea to which philosophers turned a deaf ear, but which was heard by musicians, since at Johann Sebastian Bach's death, Leibniz's De Arte Combinatoria was discovered at the composer's bedside (which, in return, permits us to read several fugues). (Serres, 1982: 46)

Seemingly describing both the real and the virtual, the infinitely large and small, the expanded and the compressed, the multiple is a model that describes the incessant ebb and flow of order and chaos that defies the imposition of a unity.

> When we subsume multiplicity under a unity, in a concept or a black box, we do not share in or parcel out information. Information is either total or null. We always assume that we don't know, or else that we know everything, yea or nay. Whereas commonly we know a bit, a meager amount, enough, quite a bit; there are various undulations, even in the hardest and most advanced sciences. (Serres, 1995: 5)

This fine balance between order and disorder (chaos) has been central to this investigation. From ecological, geopolitical, economic, and aesthetic perspectives it may never have been more pertinent. Add to this list the digitization of all aspects of life which seems paradoxically to create as much (dis)order as it does order, and we are collectively faced with some significant dilemmas and global challenges. The information revolution and its progenitors rather than installi a unified structure generate a cacophony of confusion and regret.

The preceding sections have shown how in combination different voices might harmonize to offer some solutions. Wittgenstein attempted to resolve the 'noise' of language, Serres was equally (though differently) minded to offer salvation from what he identified as a torrent of superfluous language. Where Wittgenstein falls silent in the absence of logic, Serres proposes a version of noise that is both virtual ground of potentiality and the site of over-communication as the all too literal and paradoxical attempt to impose order. Noise for Serres is a complex and mobile term vacillating between a generative field of potentiality and the bewildering presence of linguistic interventions intended to establish order – so many words, so little knowledge/wisdom. The digital then offers up a paradox – the simultaneous a-spatiality of the digital and its mathematically defined character located within a black box – exemplifies a process of expansion *and* compression. The digital is singular in its unifying of information within certain sets comprising bits, but it is also multiple and dispersed. It is conceptually distinct from the real but is also a practice *within* the real that contributes significantly to the over-production and distributed logic of language as noise.

As compelling as this undoubtedly is, what can be done with ideas like these and others that cite noise, sound, and listening as essential elements in a new mode of engagement? The aim must be to enact a process of listening to try and determine how different sensory experiences can support new modes of participation. There must be a confluence of ideas, and practice to explore new modes of engagement through a geometry of listening aligned to the articulation of spaces designated as present and visible and those that have been set to the margins (peripheral, invisible, but not silent). Such a geometry must operate as a system that is open, and which, while accommodating data lost to other modes of experiencing it, must not become mired in its own inclusivity, not drown in the tyranny of difference, but be mindful of the existence of truth, as both Serres and Badiou have urged.

New sonically inflected modes of engagement with specific spatial and temporal coordinates can facilitate much needed knowledge and understanding of our contemporary plight. Reflecting dominant thinking from the period of their construction, such coordinates that have long appeared to be frozen in time, permanent measures of cultural significance (or remnants of the past, decaying and no longer pertinent to the distribution of the sensible) and monuments to order and a time when everything stood still – can now reveal new 'insights'. Investigating the object as dataset, where sound and the diffuse temporality of noise are incorporated, serves to challenge a mode of sensory engagement with the world that has been conditioned by centuries of normative practices, resulting in a particular form of ocularcentrism (where our social, political, and aesthetic practices have in turn also been conditioned). Such practices are historically contingent, and in some ways, arbitrary (as Whitehead demonstrated in relation to the symbolism of architecture). They have plotted an anthropocentric course that is now rightly being questioned. Old certainties no longer endure, and we face a choice: to remain static and perish along with our planet and its moribund political and economic structures or embrace uncertainty and the constancy of change; experiencing the world anew, using combinations of sensory perception enabled by new technology and informed by critical thinking. This cannot wait. We must set in train immediately further research into different modes of experiencing, to challenge previously received logic and knowledge and to embrace the challenges of the future.

We must do so by embracing sound, learning what we can understand from the acoustic properties of the past,[17] and doing so in a way that, rather than clinging on to the dualism between seeing and hearing, will fuse the image and sound to form new modes of engagement and knowledge production. Despite many advances and a growing presence for sound, the visual continues to dominate. This book has tried to challenge and explore the possibilities for some of these modes of articulation and engagement. It is now essential that we test whether the addition of transient sound to an apparent permanence can alter the sensory experience of place and with it some fundamental aspects of the subject/object correlation, and from there social, political, and aesthetic motivations and ambitions and mechanisms for their realization.

Reframing subjectivity as a mode of being in the world as object with other objects – combining perception and conception to generate a field of representation where metaphysics and epistemology collide in a way that questions Heideggerian hermeneutics, we may begin to come to terms with the real as it exists outside human perception of it. Might it be paradoxically that it is at the level of the unseen (but not the unreal) that new insights lie? Engaging with things out of view, buried, hidden, forgotten, or set to the margins, either deliberately or inadvertently, or accessible via other means, maybe via listening, or by resetting our sensory relation to objects altogether, we can stimulate

[17] Christopher Witmore (2006), 'Vision, Media, Noise and the Percolation of Time: Symmetrical Approaches to the Mediation of the Material World', *Journal of Material Culture, Volume 11, issue 3.*

new ways of tackling significant contemporary challenges. Drawing on thinking – from Whitehead, Wittgenstein, through Michel Serres and up to the contemporary debates around speculative realism and object oriented ontology – it is necessary to take ideas rooted in the humanities and set them against a backdrop of technological advance to test new forms of engagement and in the process question some fundamental assumptions about humans and their place in the world. By exploring engagement with the object world and the materiality of the past and the present through new modes of sensory articulation we can better formulate our future orientation as humans in an increasingly posthuman landscape.

The digitization of the senses presents an opportunity to resolve some fundamental issues relating to our modes of experiencing the world and our access to, and grasp of, reality. Combining the visible and the audible with what might be described as an algorithmic data-real, facilitates an extra-dimensional practice prefaced on movement, diffusion, and potential rather than stasis and objective certainty. To further develop such a practice/process of engagement it is essential once more to stress Whitehead's distinction between everyday lived experience (notions) and the more systematic mode of engagement and also the recurring theme of 'the multiple' (Michel Serres' in *Genesis*) as a descriptor of both the real and the virtual. Reconciling these two modes of experiencing – the lived notional or multiple with the analytic/systematic, or unitary without recourse to privilege – means acknowledging the multiple/notion while at the same time applying systematic/unitary strategies to guard against the paradox of difference. This is important within a historical context in which humans have invariably attempted to impose some kind of order on an otherwise chaotic universe.

To be able to engage with the world as multiple requires both lived activity and analytical systems in parallel. This book has generated a survey of ideas relevant to the task at hand, and in the process of surveying has presented new ideas that can be animated through a geometry of listening. This describes an engagement that celebrates the contingency of meaning and its propensity to take multiple simultaneous forms in a way that unsettles the certainty of perception and introduces diffuse data that celebrates the potentiality of uncertainty as a reality that must be perpetually managed. This perpetual management by beings whose privileged position is critically undermined designates a new revolutionary politics of engagement.

Bibliography

Badiou, A. 2005. 'Democratic Materialism and the Materialist Dialectic', *Radical Philosophy*; https://www.radicalphilosophy.com/article/democratic-materialism-and-the-materialist-dialectic.

Balzac, H. 1988. *Gillette or The Unknown Masterpiece.* The Menard Press.

Bauman, Z. 2004. *Wasted Lives: Modernity and Its Outcasts.* Cambridge: Polity.

Borges, J. L. 2000. *Collected Fictions.* London: Penguin Classics.

Borges, J. L. 2000 *Labyrinths.* London: Penguin Classics.

Brassier, R. 2013. *Unfree Improvisation/Compulsive Freedom*; http://www.mattin.org/essays/unfree_improvisation-compulsive_freedom.html.

Brown, S. D. 2002. 'Michel Serres: Science, Translation and the Logic of the Parasite', *Theory, Culture & Society*, 19(3): 1–27.

Bryant, L., N. Srnicek, and G. Harman. 2011. *The Speculative Turn: Continental Materialism and Realism.* Melbourne: re.press.

Cox, C. 2011. 'Beyond Representation and Signification: Toward a Sonic Materialism', *Journal of Visual Culture*, 10(2): 145–61.

Deleuze, G. 1994. *Difference and Repetition.* London: Bloomsbury.

Deleuze, G., and F. Guattari. 2004a. *Anti Oedipus.* London: Bloomsbury.

Deleuze, G., and F. Guattari. 2004b. *A Thousand Plateaus.* London: Bloomsbury.

Dobereiner, L. 2014. *How to Think Sound in Itself? Towards a Materialist Dialectic of Sound.* Proceedings of the Electroacoustic Music Studies Network Conference Electroacoustic Music Beyond Performance, Berlin, June 2014.

Dobereiner 2019. 'Towards a Materialist Conception of Sound as Thing', *Organised Sound*, 24(2): 217–23.

Elfyn Jones, R. 2000. 'A. N. Whitehead and Music: Real Time', *The Musical Times*, Winter, 141(1873): 47–52.

Forrest-Kelly, T. 2014. *Capturing Music: The Story of Notation.* New York: W. W. Norton.

Foucault, M. 2001. *The Order of Things: An Archaeology of the Human Sciences.* Milton Park: Routledge.

Gaiman. N. 2007. *Fragile Things.* London: Headline Review.

Galeyev, B. M. 1993. 'Synaesthesia and Musical Space: On Yavorsky's Forgotten Hypothesis and a Proposal for an Experiment in Zero Gravity', *Leonardo*, 26(1): 73–8.

Galeyev, B. M. 2003. 'Evolution of Gravitational Synesthesia in Music: To Color and Light!', *Leonardo*, 36(2): 129–34.

Hamdan, L. A. 2021. 'The Sonic Image', *Infrasonica* https://infrasonica.org/en/audiblematter/thesonicimage.

James, W. 2000. *Pragmatism and Other Writings*. London: Penguin Classics.

Kim-Cohen, S. 2009. *In the Blink of an Ear: Toward a Non-cochlear Sonic Art*. New York: Continuum.

Kittler, F. [1986] 1999. *Gramophone, Film, Typewriter*. Stanford, CA: Stanford University Press.

Kuberski, K. 1994. *Chaosmos: Literature, Science, and Theory after Modernism*. New York. SUNY Press.

Krukowski, D. 2019. *Ways of Hearing*. Cambridge: MIT Press.

Laruelle. 2017. *Principles of Non-philosophy*. London: Bloomsbury.

Martin, F. David. 1975. *Art and the Religious Experience: The Language of the Sacred*. Lewisburg, PA: Bucknell University Press.

McHenry, L. B. 1989. 'Bradley, James, and Whitehead on Relations', *The Journal of Speculative Philosophy*, New Series, 3(3): 149–69.

McLuhan, E., and M. McLuhan. 1988. *The Laws of Media: The New Science*. Toronto: University of Toronto Press.

Nancy, J. L. 2007. *Listening*. New York: Fordham University Press.

Nietzsche, F. 1993. *The Birth of Tragedy*. Penguin Classics.

Oakeshott, M. 2015. *Experience and Its Modes*. Cambridge Philosophy Classics.

Pokorný, Vít, and Martin Nitsche. 2021. 'Philosophy and Sonic Research: Thinking with Sounds, Rhythms, and Music: An Editorial Introduction', *Open Philosophy*, 4: 374–7; https://www.degruyter.com/document/doi/10.1515/opphil-2020-0195/html?lang=en.

Ranciere, J. 2003. *The Politics of Aesthetics*. London: Bloomsbury.

James Read. 2020. 'Geometrical Constructivism and Modal Relationalism: Further Aspects of the Dynamical/Geometrical Debate. *International Studies in the Philosophy of Science*', 33(1).

Roden, D. 2014. *Posthuman Life: Philosophy at the Edge of the Human*; https://philpapers.org/rec/RODPAP-8.

Sellars, W. 1954. 'Some Reflections on Language Games', *Philosophy of Science*, 21(3) (July): 204–28.

Serres, M. 1982a. *Hermes: Literature, Science, Philosophy*, edited by J. V. Harari and D. F. Bell. Baltimore, MD: Johns Hopkins University Press.

Serres, M. 1982b. 'The Origin of Language: Biology, Information Theory and Thermodynamics', *Oxford Literary Review*, 5(1/2): 113–24.

Serres, M. 1982c. *The Parasite*. Baltimore, MD: Johns Hopkins University Press.

Serres, M. 1995. *Genesis*, translated by G. James and J. Nielson. Ann Arbor: University of Michigan Press.

Serres, M. 2008. *The Five Senses: A Philosophy of Mingled Bodies*. London: Bloomsbury.

Serres, M. 2017. *Geometry: The Third Book of Foundations*. London: Bloomsbury.

Scruton, R. 1981. *Kant: A Very Short Introduction*. Oxford: Oxford University Press.

Shaviro, S. 2009. *Without Criteria: Kant, Whitehead Deleuze and Aesthetics*. Cambridge: MIT Press.

Stengers, I. 2014. *Thinking with Whitehead: A Free and Wild Creation of Concepts*. Cambridge, MA: Harvard University Press.

Sterne, J. 2006. 'The mp3 as Cultural Artifact', *New Media & Society*, 8(5): 825–42.

Sterne, J. 2012. *The Meaning of a Format: Sign, Storage, Transmission*. Durham, NC: Duke University Press.

Sterret, S. 2005. 'Pictures of Sounds: Wittgenstein on Gramophone Records and the Logic of Depiction', *Studies in History and Philosophy of Science*, 36A(2): 351–62.

Tymoczko, Dmitri. 2011. *A Geometry of Music: Harmony and Counterpoint in the Extended Common Practice*. Oxford Studies in Music Theory. Oxford: Oxford University Press.

Whitehead, A. N. 1920. *The Concept of Nature*. Cambridge: Cambridge University Press.

Whitehead, A. N. 1968. *Modes of Thought*. New York: Free Press.

Whitehead, A. N. 1969. *Process & Reality*. New York: Free Press.

Whitehead, A. N. 1985. *Symbolism: Its Meaning and Effect*. New York: Fordham University Press.

Wittgenstein, L. 2009. *Tractus Logico-Philosophicus*. Milton Park: Routledge.

Zellini, P. 2020. *The Mathematics of the Gods and the Algorithms of Men: A Cultural History*. London: Allen Lane.

Zuckerkandl, Victor. 1973. *Sound and Symbol: Music and the External World*. Princeton, NJ: Princeton University Press.

Index

www.ingramcontent.com/pod-product-compliance
Ingram Content Group UK Ltd.
Pitfield, Milton Keynes, MK11 3LW, UK
UKHW020208290926
15470UKWH00001B/75

9 781501 369346